The Strange Ca the Spotted Mice

and other classic essays on science

Sir Peter Medawar, OM, 1915–87, was born in Rio de Janeiro and educated at Magdalen College, Oxford. He began research in H. W. Florey's department at Oxford in the early days of the development of penicillin. After professorships at Birmingham and University College London, he became Director of the National Institute for Medical Research. His scientific reputation is based mainly on his research in immunology, which helped make transplant surgery possible. In 1960 he won the Nobel Prize for Medicine for his work on tissue transplantation. Elected to the Royal Society at the age of 34, he was also a Fellow of the British Academy—a rare honour for a scientist.

Sir Peter wrote a number of books for a general audience, including *Pluto's Republic* (1982), *Aristotle to Zoos* (1983, with Jean Medawar), and *The Limits of Science* (1985). A further collection of his essays, *The Threat and the Glory* (1990), was published after his death.

The Strange Case of the Spotted Mice

and other classic essays on science

Peter Medawar

Foreword by
Stephen Jay Gould

Oxford   New York
OXFORD UNIVERSITY PRESS

Oxford University Press, Walton Street, Oxford OX2 6DP

Oxford New York
Athens Auckland Bangkok Bombay
Calcutta Cape Town Dar es Salaam Delhi
Florence Hong Kong Istanbul Karachi
Kuala Lumpur Madras Madrid Melbourne
Mexico City Nairobi Paris Singapore
Taipei Tokyo Toronto
and associated companies in
Berlin Ibadan

Oxford is a trade mark of Oxford University Press

First published as an Oxford University Press paperback 1996

British Library Cataloguing in Publication Data
Data available

Library of Congress Cataloging in Publication Data
Data available
ISBN 0–19–286193–X

10 9 8 7 6 5

Typeset by Best-set Typesetter Ltd., Hong Kong
Printed in Great Britain by Mackays
Chatham, Kent

Foreword

The Phenomenon of Medawar
Stephen Jay Gould

Peter Medawar was a paragon of rationalism. I have never met a man more committed to that combination of logic and common sense that we call science at its best. I have also never met a tougher or more confident man. Reason and fortitude forge an unbeatable combination. One personal story: I attended a scientific meeting with Peter Medawar after a stroke had made half his body virtually unusable. We had to move from the third-floor lecture room to the basement restaurant—and the building had no elevator. Peter could only go down stairs backwards, and slowly. Most people in his condition would have meekly waited for everyone else to descend and then painfully made their own way down, no doubt missing half the lunch by late arrival. Peter was the star of the meeting—and he knew it. He went down first, one step at a time, slowly as could be—and everyone followed at his pace. How entirely right and proper, we all agreed.

Peter Medawar was also a paragon of humanism—and this form of compassion took the edge off the occasional harshness of his rationalism. He loved people and their foibles, and he loved to laugh. He was a true philosopher, but he also reminded me of the fellow in Johnson's famous quip about the man who wanted to be a philosopher, but failed because cheerfulness was always breaking through. Another personal story: A year or two after the British forum with the luncheon descent, Peter and I attended another meeting in Minnesota. He had suffered another stroke, and was now confined to a wheelchair. Meanwhile, I had come down with an apparently incurable cancer. Peter was my guru and I wanted, more than anything else at the meeting, to ask him as an immunologist by trade, and especially in the light of his own severe problems of health, how and whether, in his opinion, attitude might help in the palliation or remediation of physical illness. So I asked him what traits of character or action might help and he replied simply: 'a sanguine personality'. We both benefited by his judgment, having been endowed for whatever reason with such a view of the world. I recovered, and he lived far longer (and better) with half a body than the vast majority of people could ever hope to survive with all systems functional.

This toughness, combined with compassion and good humor, represents the persona that a commitment to rationalism is supposed to induce, but so rarely does (for personal philosophy does not always overcome personal insecurity). The dispelling of illusions, and acceptance of the world as we actually find it (rather than as we could wish it to be) should be profoundly liberating, though not always joyous. These views underlie the great Spinozistic notion—the deep meaning of true liberty to a philosophic determinist (as the extreme), or simply to one who does not deny the constraints of objective reality—that 'freedom is the recognition of necessity'. With such acceptance, one can be truly sanguine and accept all the outrageous slings and arrows of life's misfortune with both good grace and good humour. A fine idea—although most people don't have the internal strength to embody the consistent philosophy. But Peter Medawar did. Let him speak for himself:

What matters is not to be defeated. I do not regard myself either as a victim or a beneficiary of divine dispensations, and I do not believe— much though I should like to do so—that God watches over the welfare of small children in the way that small children need looking after (that is, as fond parents do, and paediatricians, and good schoolteachers). I do not believe that God does so because there is no reason to believe it. I suppose that's just my trouble: always wanting reasons.

In a world awash in new-age mysticism and old-fashioned, pervasive anti-intellectualism, it may be something of a cliché

to point out that salvation requires rationalism now more than ever. Equally obviously, such salutary attitudes need to be embodied in passionate and persuasive individuals, not only in abstractions or collectivities. In this sense, although independently minded intellectuals like myself are not supposed to have gurus, I don't mind admitting that Peter Medawar was, for me, closest to that status of more than mere mentor. I was not blind to his faults (and will mention some later on), but I never met more qualities that I wanted to emulate all rolled up into one person. It goes without saying that Peter Medawar, Nobel laureate for work in immunology, was a pre-eminent biologist. But he was also a great intellectual and committed rationalist, a firm moralist with a redeeming sense of humour, and a wonderful writer especially gifted with the skill to turn a fine phrase. These combined qualities, so rarely conjoined in a single individual, made Medawar our century's greatest spokesman for the power and humanity of science. Just consider a pair of his *bon mots*: first, an incisive comment on psychobabble in the search for intrinsic meaning in dreams:

those who enjoy slopping around in the amniotic fluid should pause for a moment to entertain (perhaps only unconsciously in the first instance) the idea that the content of dreams may be totally devoid of 'meaning'.

Or this, in a kinder mode, on the difference between art and science:

Wagner would certainly not have spent twenty years on *The Ring* if he had thought it at all possible for someone else to nip in ahead of him with *Götterdämmerung*.

I have frequently argued (see my books *Bully for Brontosaurus* and *Eight Little Piggies*) that good popular writing in science should be divided into two basic modes: Franciscan (to honour a great poet; and a lover of nature as well as God) for lyrical writing about nature's loveliness, and Galilean (to honour a great rationalist who wrote his books as Italian dialogues for all to read, rather than as Latin treatises for the few) for the joy of intellectual resolution of nature's numerous puzzles. In

this dichotomy, Peter Medawar is (along with J. B. S. Haldane) the Galilean hero of the twentieth century, just as T. H. Huxley wins the palm for the nineteenth.

Medawar followed the grand lineage of Montaigne and used the essay as his weapon of wit and instruction. So well did he ply this trade that even his book reviews, usually the prime example of an ephemeral genre never worth republishing by definition, became wise essays on enduring themes, with the particular book just used as an entrée to the generality (as several chapters in this collection will show).

Most of Medawar's essays follow the classical pattern of sceptical rationalism in a world of unreason—use of a logical weapon against a series of targets. In the largest sense, Medawar's weapon is science in general—and he does recognize the multifarious nature of valid scientific methodologies. More particularly, Medawar was an uncritical disciple of Sir Karl Popper, and his arguments hew strictly to the Popperian doctrine of falsificationism (statements cannot be proven absolutely true, but can be conclusively falsified; statements not subject to falsification in principle are not scientific). I find Popperianism narrow in some ways, outdated in others, and in this sense cannot agree with all of Medawar's methodology. If Medawar was my guru, then Popper was certainly Medawar's—and perhaps we only learn from this that intellectuals really shouldn't have gurus. His targets are the enemies of science and reason—particularly those who become cult figures and threaten the rationalist perspective thereby. Medawar had little use for mystics (and their unfalsifiable statements) masquerading as scientists—and his dismissal of Teilhard de Chardin, when the Jesuit palaeontologist became a cult figure in the 1960s for his woolly, mystical (and false) version of evolution, has rightly become one of the great essays of our century, properly given pride of initial place in this volume. With his rapier, Medawar showed that Teilhard's cult classic, *The Phenomenon of Man*, is conventional clap-trap mysticism, pretending to be deep and original because the opacity of Teilhard's writing leads the unsuspecting to regard him as profound:

the greater part of it, I shall show, is nonsense, tricked out with a variety of metaphysical conceits, and its author can be excused of dishonesty only on the grounds that before deceiving others he has taken great pains to deceive himself. . . . it is the style that creates the illusion of content, and . . . is a cause as well as merely a symptom of Teilhard's alarming apocalyptic seizures.

Freudian psychoanalysis receives treatment almost as dismissive—with a passion that rings a bit archaically in our decidedly post-Freudian age, though we should remember what a hold (and at what expense) this undocumented theory (also undocumentable in Popper's sense) had upon so many people just a few decades ago. Medawar summarizes his complaint with ample justice:

The property that gave psychoanalysis the character of a mythology is its combination of conceptual barrenness with an enormous facility of explanation.

Medawar was a great humanist himself, but his third major target consisted of pompous professionals in the humanities, who assumed that their enterprise represented the sole height of human achievement and decency, and that science could be dismissed as a form of join-the-dots engineering. (Medawar often tellingly remarked that a scientist would be considered ignorant—and properly so, Medawar agreed—if he knew nothing of Shakespeare or Beethoven, but that humanists often took pride in knowing equally little about Darwin or Newton.) Here, I think, Medawar did sometimes go overboard (although I well understand the provocation), and could become both ungenerous and more than mildly élitist in the pass-the-port-this-way style of high-table Oxbridge (a trait that can never endear a patrician Englishman to a Yank like me). I do regard this passage about humanists who liked Teilhard as doing more harm than good for the cause that both Medawar and I share:

How have people come to be taken in by *The Phenomenon of Man?* We must not underestimate the size of the market for works of this kind, for philosophy-fiction. Just as compulsory primary education created a market catered for by cheap dailies and weeklies, so the spread of secondary and latterly of tertiary education has created a large population of people,

often with well-developed literary and scholarly tastes, who have been educated far beyond their capacity to undertake analytical thought.

Medawar's power of prose consisted largely in his remarkable ability to combine toughness, beauty and clarity (all with a thorough lacing of wit). Any of his essays will do, but let me illustrate this winning conjunction with some statements from a short, and otherwise minor, piece on Florey's role in the discovery of penicillin. This passage for toughness and brevity of wit:

The mice were watched all night (but of course). All four mice unprotected by penicillin had died by 3:30 a.m. Heatley recorded the details and cycled home in the black-out. Poor mice? Yes of course poor mice, but poor human beings too, don't forget.

And this for passionate involvement, and quality of prose:

Macfarlane's account of the animal experiments and the first clinical trial is simple and straightforward, and all the more exciting for being so. It makes my heart pound still although I know the outcome, for the thrill of reading about these great occasions does not diminish . . .

And finally, for a gem of clear explanation (and for a difficult concept):

The widespread use of penicillin—sometimes—injudiciously often—had led to the evolution in many hospitals of strains of bacteria resistant to its action: once a mutant impervious to penicillin has arisen, natural selection soon brings it about that the mutant becomes the prevailing type in the population. It is not that penicillin has lost any virtue, but rather that bacteria have acquired a vice.

People sometimes make the mistake of seeing essays like these—most in the debunking mode of exploding myths of unreason—as negative exercises, however useful. They may then make the further error of assuming that anything negative must be constricting in scope and fundamentally small-minded by definition. No misconception could be deeper. Walt Whitman, America's greatest nineteenth-century poet, urged us to 'make much of negatives'. Debunking can be purely subtractive (and still fun if done with wit and a rapier), but such is never Medawar's way, nor the manner of any great thinker in this venerable tradition. One does not debunk in order to demolish alone, but

rather because an existing edifice seems harmful (or at least seriously in the way of a grand view). One debunks, in other words, in the interest of an alternative view of life. We need the uncompromising humanistic rationalism of Peter Medawar, now more than ever. I shall let Peter Medawar speak for himself in closing, in a fine passage about the narcissistic, and ultimately anti-humanistic, content of modern psychobabble masquerading as philosophy:

There is a particular selfishness about modern philosophic speculation (using 'philosophy' here again in its homely or domestic sense). The philosophic universe has contracted into a neighborhood, a suburbia of personal relationships. It is as if the classical formula of self-interest, 'I'm all right, Jack' was seeking a new context in our private, inner world. We can, obviously, do better than this.

Museum of Comparative Zoology
Harvard University
Cambridge, MA 02138

Contents

Introduction
David Pyke

Sir Peter Medawar was three great men. He was a great scientist, a man of great courage—and a great writer. He was supremely creative both as a scientist and as a writer, defining creativity as 'the faculty of mind or spirit that empowers us to bring into existence, ostensibly out of nothing, something of beauty, order or significance'. His creativity in literature was shown in his volumes of essays, especially in *The Art of the Soluble* and *The Hope of Progress*. They consisted largely of reviews, talks, or lectures and, though they were about science and scientists, were written for a general audience.

If I seem to labour the point by saying that he was as great a writer as a scientist it is partly because I agree with him that 'a man's style of writing is an important part of his character—some would say one of the most revealing parts'. His recipe for good writing was this: 'Brevity, cogency and clarity are the principal virtues and the greatest of these is clarity.'

Peter was born in Brazil in 1915 of a Lebanese father and English mother. He was sent to school in England and lived there for the rest of his life. When he was still at preparatory school he realized that he was 'hooked on science; no other kind of life would do'. He went to Marlborough, then Magdalen College, Oxford. He got a first class degree in Zoology and then became a Research Fellow.

After exploring various lines of research he focused on the problem of why skin is rejected when grafted from one person to another. He showed that the rejection of skin, kidney, or any other organ is under immunological control. Previously rejection had been considered genetic in origin and therefore insurmountable. But after five years' work he demonstrated, in a series of brilliant experiments, that the barrier could be overcome. The importance of the discovery, which depends upon grafting 'foreign' cells into an animal while it is still *in utero*, was immediately appreciated and led to the award of the Nobel Prize in 1960. It gave great encouragement to the whole medical scientific community and created the new speciality of transplant surgery. Immunologists from all over the world came to work with him and any budding transplant surgeon hoped to have the chance to do so. Peter's discovery of 'immunological tolerance' was important not because it showed *how* rejection by one person of tissue from another could be overcome but *that* it could be overcome. Today's techniques of immune suppression use complex drugs; they would not have been discovered—or not discovered as soon as they were—without Peter's work. He showed the way.

Peter had written about science and scientists for a general audience in *The Uniqueness of the Individual* in 1957 and in the BBC Reith Lectures *The Future of Man* in 1959. After the Nobel Prize in 1960 he was asked to speak, review, and broadcast all over the world. He had a great drive to convey to others the meaning and importance of science and to explain his own passion for it. Unlike most scientists he was interested in the philosophy of science, the process of scientific discovery. Science progresses by imaginative leaps which are put to the test in the laboratory: if they fail the test the ideas are discarded, if they pass they survive until later experiments refute or modify them. But what produces the ideas in the first place? They do not enter a vacant mind nor emerge from a heap of randomly assembled observations. 'The most interesting and exciting of all intellectual problems is *how* the imagination is harnessed for the performance of scientific work, so that the steam, instead of blowing off in picturesque clouds and rattling the lid of the kettle is now made to turn a wheel.'

Peter was consumed by science, 'incomparably the most successful activity human beings have ever engaged upon'. It gave him no rest. Scientists are often thought of as problem-solvers. But they are more than that, they are problem-seeking creatures too. 'The motive force that is behind the scientist and technologist's almost compulsive desire for an understanding and mastery of nature is sometimes described as "curiosity" or "inquisitiveness", but these nursery words hardly do justice to what feels like a deep-seated biological impulse—the hunting feeling, I call it myself.'

Some non-scientists, and perhaps scientists, look with alarm at the progress of science, at the enormous accumulation of facts. They fear that we shall all be overwhelmed by the sheer size of the store of information. On the contrary 'the ballast of factual information, so far from being just about to sink us, is growing daily less . . . In all sciences we are being progressively relieved of the burden of singular instances, the tyranny of the particular. We need no longer record the fall of every apple.'

Peter's writings became famous. One of the most famous was his devastating review of Teilhard de Chardin's *The Phenomenon of Man*. Teilhard was a French priest who claimed to have proved the existence of God scientifically and was widely acclaimed for having done so. But his style was, to put it mildly, opaque. Peter hated opacity in thought or writing. 'People who write obscurely are either unskilled in writing or up to some mischief.' Peter wrote of Teilhard's 'tipsy, euphoristic prose-poetry' in which 'a feeble argument was abominably expressed'. 'It is written in an all but totally unintelligible style, and this is construed as *prima facie* evidence of profundity.' He saw similar pretensions and obscurity in some of the writings of psychoanalysts, whom he condemned for 'the Olympian glibness of their thought, their complete lack of hesitancy and bewilderment in the face of enormously difficult problems. On the contrary a lava-flow of *ad hoc* explanations pours over and around all difficulties, leaving only a few smoothly rounded prominences to mark where they might have lain.'

Much of Peter's writing was light-hearted, but all was serious. Nowhere is this better shown than in the last lines of his review of J. D. Watson's book *The Double Helix*. Watson was the codiscoverer of the structure of DNA, the most important biological discovery of the twentieth century, which explains how characteristics are passed from one generation to the next. His book is amusing, irreverent, naïve, revealing, and vulgar. 'The characters in the discovery of DNA come out larger than life, perhaps, and as different one from another as Caterpillar and Mad Hatter. Watson's childlike vision makes them seem like the creatures of a Wonderland, all at a strange contentious noisy tea party which made room for him because for people like him, at this particular kind of party, there is always room.' Those last fourteen words contain Peter's real message.

Watson was accused of excessive ambition and hunger for priority. Peter saw it differently. A scientist's discovery, unlike an artist's creation, is *his* only in the sense that he made it first. Peter described the difference between artistic and scientific priority in a letter declining to take part in a BBC radio game of scientists versus artists. 'Darwin's claim could not be defended against Beethoven's, even in fun . . . I don't think any scientist can be defended against a major artist; scientists are always dispensable, for, in the long run, others will do what they have been unable to do themselves.'

In 1962 Peter became Director of the National Institute for Medical Research at Mill Hill in London. He set himself to master the job and get to know all the 200 staff at the Institute. He administered the business in three and a half days a week and kept Tuesdays and Thursdays for his own research. Under his direction, the National Institute became a world leader in immunology.

It was as certain as anything can be that he would have been elected President of the Royal Society. This was perhaps the one honour he ever coveted—head of his profession and heir of Isaac Newton. He was riding the crest of a very large wave. There seemed no reason why it should not go on and on.

The wave broke on Sunday, 6 September 1969. Peter, as president, was reading the lesson in Exeter Cathedral at the annual service of the British Association for the Advancement of Science when his voice started to falter. He was helped back to his seat and there sank into unconsciousness. He had had a large cerebral haemorrhage. He was in hospital for months and was near to death twice. While still very ill and when no one knew how badly his mind had been affected he joked about his reading of the lesson in Exeter Cathedral: 'People simply don't know the risks they run when they meddle with the supernatural.' He knew how ill he had been: his reaction was characteristic: 'I myself, naturally sanguine, had considered and dismissed the possibility of dying.'

His experiences in hospital and his close encounters with death did not cause him to bother about the 'dignity' of dying. As he wrote years later: 'No thought of dignity entered my head—it is a state of mind not easily compatible with the hospital microcosm of bedpans and catheters. I needed all the help I could get to promote my ambition to remain alive. It was as allies, then, that I regarded my physicians and the apparatus of intensive care and not as so many plots to deprive me of my dignity.'

His sanguine attitude to life put him out of sympathy with people such as Ivan Illich and Thomas McKeown* who ascribe all improvements in health over the last century to social and economic factors and dismiss the advances of clinical medicine. 'So long as human beings retain their strong preference for being alive as opposed to being dead so long will medical treatment, if necessary of a strenuous and heroic character, remain in demand. I myself should rather not have needed treatment, but as I did need it, thank God I got it.' Peter got the treatment all right but for the rest of his life he was paralysed in the left arm and leg and could not see anything on the left side of his visual field.

He went back to work in 1970, although not with quite the same ferocious intensity as before. He could not use his left hand, so he could no longer do animal experiments. He started to travel again, always with his wife Jean who, as he put it, was his third leg and arm and eye. His enormous scientific output fell after 1969, but his literary output did not. He wrote seven books (not counting *The Hope of Progress*, published in 1972 but consisting of essays written before 1969). He also gave lectures and radio talks and wrote many reviews.

In 1980, while in New York, Peter had another stroke in the mid-brain which affected speech, swallowing, and walking. Again there was gradual recovery but it was not complete. He went on

* See Chapter 15.

working, writing, speaking, and travelling; supported by Jean, making little of his difficulties.

What a figure he was—6'4", still extremely handsome, half blind, walking with a stick and caliper, his left arm in a sling, his speech a little indistinct but punctuated by laughter. His physical problems were ignored. He was 'heroic in his indifference to increasing disability'. Nothing seemed to impair his enjoyment of life. His enjoyment was infectious. He loved to have company, and he was far more sociable than he had been before 1969 when he could seem austere and formidable. His friends loved to be with him. There was none of the reluctance or embarrassment one sometimes feels on being with an old friend who is so afflicted; he accepted the situation as it was and as he knew it would remain—until it got worse. It did get worse, in June 1985, a few days after he had finished his autobiographical *Memoir of a Thinking Radish*. He had several more stokes until at last, on 2 October 1987, at the age of seventy-two, he died.

I had the great good fortune to know Peter well. We first met in 1959 when his wife Jean joined me in editing a journal on family planning. After his stroke in 1969 when he spent more time at home I came to know him better and to admire him even more. I was once asked if I thought his stroke had affected his intelligence. I said I thought it had: it had reduced his IQ to three figures.

I want to end with Peter's description of the qualities needed by a scientist: 'A sanguine temperament that expects to be able to solve a problem; power of application and that kind of fortitude that keeps them erect in the face of much that might otherwise cast them down; and above all, persistence, a refusal bordering upon obstinacy to give up and admit defeat.'

Peter Medawar wrote that about scientists. It could have been written about himself.

Royal College of Physicians
London

1 The Phenomenon of Man

Everything does not happen continuously at any one moment in the universe. Neither does everything happen everywhere in it.

There are no summits without abysses.

When the end of the world is mentioned, the idea that leaps into our minds is always one of catastrophe.

Life was born and propagates itself on the earth as a solitary pulsation.

In the last analysis the best guarantee that a thing should happen is that it appears to us as vitally necessary.

This little bouquet of aphorisms, each one thought sufficiently important by its author to deserve a paragraph to itself, is taken from Père Teilhard's *The Phenomenon of Man*.[1] It is a book widely held to be of the utmost profundity and significance; it created something like a sensation upon its publication in France, and some reviewers hereabouts called it the Book of the Year—one, the Book of the Century. Yet the greater part of it, I shall show, is nonsense, tricked out with a variety of metaphysical conceits, and its author can be excused of dishonesty only on the grounds that before deceiving others he has taken great pains to deceive himself. *The Phenomenon of Man* cannot be read without a feeling of suffocation, a gasping and flailing around for sense. There is an argument in it, to be sure—a feeble argument, abominably expressed—and this I shall expound in due course; but consider first the style, because it is the style that creates the illusion of content, and which is a cause as well as merely a symptom of Teilhard's alarming apocalyptic seizures.

The Phenomenon of Man stands square in the tradition of *Naturphilosophie*, a philosophical indoor pastime of German origin which does not seem even by accident (though there is a great deal of it) to have contributed anything of permanent value to the storehouse of human thought. French is not a language that lends itself naturally to the opaque and ponderous idiom of nature-philosophy, and Teilhard has accordingly resorted to the use of that tipsy, euphoristic prose-poetry which is one of the more tiresome manifestations of the French spirit. It is of the nature of reproduction that progeny should outnumber parents, and of Mendelian heredity that the inborn endowments of the parents should be variously recombined and reassorted among their offspring, so enlarging the population's candidature for evolutionary change. Teilhard puts the matter thus: it is one of his more lucid passages, and Mr Wall's translation, here as almost everywhere else, captures the spirit and sense of the original.

Reproduction doubles the mother cell. Thus, by a mechanism which is the inverse of chemical disintegration, *it multiplies without crumbling*. At the same time, however, it transforms what was only intended to be prolonged. Closed in on itself, the living element reaches more or less quickly a state of immobility. It becomes stuck and coagulated in its evolution. Then by the act of reproduction it regains the faculty for inner re-adjustment and consequently takes on a new appearance and direction. The process is one of pluralization in form as well as in number. The elemental ripple of life that emerges from each individual unit does not spread outwards in a monotonous circle formed of individual units exactly like itself. It is diffracted and becomes iridescent, with an indefinite scale of variegated tonalities. The living unit is a centre of irresistible multiplication, and *ipso facto* an equally irresistible focus of diversification.

In no sense other than an utterly trivial one is reproduction the inverse of chemical disintegration. It is a misunderstanding of genetics to suppose that reproduction is only 'intended' to make facsimiles, for parasexual processes of genetical exchange are to be found in the simplest living things. There seems to be some confusion between the versatility of a population and the adaptability of an individual. But errors of fact or judgement of this kind are to be found throughout, and are not my immediate concern; notice instead the use of adjectives of excess (misuse, rather, for genetic diversity is not indefinite nor multiplication irresistible). Teilhard is for ever shouting at us: things or affairs are, in alphabetical order, astounding, colossal, endless, enormous, fantastic, giddy, hyper-, immense, implacable, indefinite, inexhaustible, inextricable, infinite, infinitesimal, innumerable, irresistible, measureless, mega-, monstrous, mysterious, prodigious, relentless, super-, ultra-, unbelievable, unbridled or unparalleled. When something is described as merely *huge* we feel let down. After this softening-up process we are ready to take delivery of the neologisms: biota, noosphere, hominization, complexification. There is much else in the literary idiom of nature-philosophy: *nothing-buttery*, for example, always part of the minor symptomatology of the bogus. 'Love in all its subtleties is nothing more, and nothing less, than the more or less direct trace marked on the heart of the element by the psychical convergence of the universe upon itself.' 'Man discovers that he is *nothing else than evolution become conscious of itself,*' and evolution is 'nothing else than the continual growth of . . . "psychic" or "radial" energy'. Again, 'the Christogenesis of St Paul and St John is nothing else and nothing less than the extension . . . of that noogenesis in which cosmogenesis . . . culminates'. It would have been a great disappointment to me if Vibration did not somewhere make itself felt, for all scientistic mystics either vibrate in person or find themselves resonant with cosmic vibrations; but I am happy to say that on page 266 Teilhard will be found to do so.

These are trivialities, revealing though they are, and perhaps I make too much of them. The evolutionary origins of consciousness are indeed distant and obscure, and perhaps so trite a thought does need this kind of dressing to make it palatable: 'refracted rearwards along the course of evolution, consciousness displays itself qualitatively as a spectrum of shifting hints whose lower terms are lost in the night' (the roman type is mine). What is much more serious is the fact that Teilhard habitually and systematically cheats with words. His work, he has assured us, is to be read, not as a metaphysical system, but 'purely and simply as a scientific treatise' executed with 'remorseless' or 'inescapable' logic; yet he uses in metaphor words like energy, tension, force, impetus and dimension *as if* they retained the weight and thrust of their special scientific usages. Consciousness, for example, is a matter upon which Teilhard has been said to have illuminating views. For the most part consciousness is treated as a manifestation of energy, though this does not help us very much because the word 'energy' is itself debauched; but elsewhere we learn that consciousness is a dimension, or is something with mass, or is something corpuscular and particulate which can exist in various degrees of concentration, being sometimes infinitely diffuse. In his lay capacity Teilhard, a naturalist, practised a comparatively humble and unexacting kind of science, but he must have known better than to play such tricks as these. On page 60 we read:

The simplest form of protoplasm is already a substance of unheard-of complexity. This complexity increases in geometrical progression as we pass from the protozoon higher and higher up the scale of the metazoa. And so it is for the whole of the remainder always and everywhere.

Later we are told that the *'nascent* cellular world shows itself to be already infinitely complex'. This seems to leave little room for improvement. In any event complexity (a subject on which Teilhard has a great deal to say) is not measurable in those scalar quantities to which the concept of a geometrical progression applies.

In spite of all the obstacles that Teilhard perhaps wisely puts in our way, it is possible to discern a train of thought in *The Phenomenon of Man*. It is founded upon the belief that the fundamental process or motion in the entire universe is *evolution*, and evolution is 'a general condition to which all theories, all hypotheses, all systems must bow . . . a light illuminating all facts, a curve that all lines must follow'. This being so, it follows that 'nothing could ever burst forth as final across the different thresholds successively traversed by evolution . . . which has not already existed in an obscure and primordial way' (again my romans). Nothing is wholly new: there is always some primordium or rudiment or archetype of whatever exists or has existed. Love, for example—'that is to say, the affinity of being with being'—is to be found in some form throughout the organic world, and even at a 'prodigiously rudimentary level', for if there were no such affinity between atoms when they unite into molecules it would be 'physically impossible for love to appear higher up, with us, in "hominized" form'. But above all, consciousness is not new, for this would contradict the evolutionary axiom; on the contrary, we are 'logically forced to assume the existence in rudimentary form . . . of some sort of psyche in every corpuscle', even in molecules; 'by the very fact of the individualization of our planet, a certain mass of elementary consciousness was originally imprisoned in the matter of earth'.

What form does this elementary consciousness take? Scientists have not been able to spot it, for they are shallow superficial fellows, unable to see into the inwardness of things—'up to now, has science ever troubled to look at the world other than from *without*?' Consciousness is an interiority of matter, an 'inner face that everywhere duplicates the "material" external face, which alone is commonly considered by science'. To grasp the nature of the within of things we must understand that energy is of two kinds: the 'tangential', which is energy as scientists use that word, and a radial energy (a term used interchangeably with spiritual or psychic energy), of which consciousness is treated sometimes as the equivalent, sometimes as the manifestation, and sometimes as the consequence (there is no knowing what Teilhard intends). Radial energy appears to be a measure of, or that which conduces towards, complexity or degree of arrangement; thus 'spiritual energy, by its very nature, increases in "radial" value . . . in step with the increasing chemical complexity of the elements of which it represents the inner lining'. It confers *centricity*, and 'the increase of the synthetic state of matter involves . . . an increase of consciousness'.

We are now therefore in a position to understand what evolution is (is nothing but). Evolution is 'the continual growth of . . . "psychic" or "radial" energy, in the course of duration, beneath and within the mechanical energy I called "tangential" '; evolution, then, is 'an ascent towards consciousness'. It follows that evolution must have a 'precise *orientation* and a privileged *axis*' at the topmost pole of which lies Man, born 'a direct lineal descendant from a total effort of life'.

Let us fill in the intermediate stages. Teilhard, with a penetrating insight that Sir Julian Huxley singles out for special praise, discerns that consciousness in the everyday sense is somehow associated with the possession of nervous systems and brains ('we have every reason to think that in animals too a certain inward-ness exists, approximately proportional to the development of their brains'). The direction of evolution must therefore be towards cerebralization, that is, towards becoming brainier. 'Among the infinite modalities in which the complication of life is dispersed,' he tells us, 'the differentiation of nervous tissue stands out . . . as a significant transformation. *It provides a direction*; and by its consequences *it proves that evolution has a direction*.' All else is equivocal and insignificant; in the process of becoming brainier we find 'the very essence of complexity, of essential metamorphosis'. And if we study the evolution of living things, organic evolution, we shall find that in every one of its lines, except only in those in which it does not occur, evolution is an evolution towards increasing complexity of the nervous system and cerebralization. Plants don't count, to be sure (because 'in the vegetable kingdom we are unable to follow along a nervous system the evolution of a psychism obviously remaining diffuse'), and the contemplation of insects provokes a certain shuffling of the feet;[2] but primates are 'a phylum of *pure and direct cerebralization*' and among them 'evolution went straight to work on the brain, neglecting everything else'. Here is Teilhard's description of noogenesis, the birth of higher consciousness among the primates, and of the noosphere in which that higher consciousness is deployed:

By the end of the Tertiary era, the psychical temperature in the cellular world had been rising for more than 500 million years . . . When the anthropoid, so to speak, had been brought 'mentally' to boiling-point some further calories were added . . . No more was needed for the whole inner equilibrium to be upset . . . By a tiny 'tangential' increase, the 'radial' was turned back on itself and so to speak took an infinite leap forward. Outwardly, almost nothing in the organs had changed. But in depth, a great revolution had taken place: consciousness was now leaping and boiling in a space of super-sensory relationships and represen-tations . . .

The analogy, it should be explained, is with the vaporization of water when it is brought to boiling-point, and the image of hot vapour remains when all else is forgotten.

I do not propose to criticize the fatuous argument I have just outlined; here, to expound is to expose. What Teilhard seems to be trying to say is that evolution is often (he says always) accompanied by an increase of orderliness or internal coherence or degree of integration. In what sense is the fertilized egg that develops into an adult human being 'higher' than, say, a bacterial cell? In the sense that it contains richer and more complicated genetical instructions for the execution of those processes that together constitute development. Thus Teilhard's radial, spiritual or psychic energy may be equated to 'information' or 'information content' in the sense that has been made reasonably precise by modern communications engineers. To equate it to consciousness, or to regard degree of consciousness as a measure of information content, is one of the silly little metaphysical conceits I mentioned in an earlier paragraph. Teilhard's belief, enthusiastically shared by Sir Julian Huxley, that evolution flouts or foils the second law of thermodynamics is based on a confusion of thought; and the idea that evolution has a main track or privileged axis is unsupported by scientific evidence.

Teilhard is widely believed to have rejected the modern Mendelian-Darwinian theory of evolution or to have demonstrated its inadequacy. Certainly he imports a ghost, the entelechy or *élan vital* of an earlier terminology, into the Mendelian machine; but he seems to accept the idea that evolution is probationary and exploratory and mediated through a selective process, a 'groping', a 'billionfold trial and error'; 'far be it from me', he declares, 'to deny its importance'. Unhappily Teilhard has no grasp of the real weakness of modern evolutionary theory, namely its lack of a complete theory of variation, of the origin of *candidature* for evolution. It is not enough to say that 'mutation' is ultimately the source of all genetical diversity, for that is merely to give the phenomenon a name: mutation is so defined. What we want, and are very slowly beginning to get, is a comprehensive theory of the forms in which new genetical information comes into being. It may, as I have hinted elsewhere, turn out to be of the nature of nucleic acids and the chromosomal apparatus that they tend spontaneously to proffer genetical variants—genetical solutions of the problem of remaining alive—which are more complex and more elaborate than the immediate occasion calls for; but to construe this 'complexification' as a manifestation of consciousness is a wilful abuse of words.

Teilhard's metaphysical argument begins where the scientific argument leaves off, and the gist of it is extremely simple. Inasmuch as evolution is the fundamental motion of the entire universe, an ascent along a privileged and necessary pathway towards consciousness, so it follows that our present consciousness must 'culminate forwards in some sort of supreme consciousness'. In expounding this thesis, Teilhard becomes more and more confused and excited and finally almost hysterical. The Supreme Consciousness, which apparently assimilates to itself all our personal consciousnesses, is, or is embodied in, 'Omega' or the Omega-point; in Omega 'the movement of synthesis culminates'. Now Omega is 'already in existence and operative at the very core of the thinking mass', so if we have our wits about us we should at this moment be able to detect Omega as 'some excess of personal, extra-human energy', the more detailed contemplation of which will disclose the Great Presence. Although already in existence, Omega is added to progressively: 'All round us, one by one, like a continual exhalation, "souls" break away, carrying upwards their incommunicable load of consciousness', and so we end up with 'a harmonized collectivity of consciousnesses equivalent to a sort of super-consciousness'.

Teilhard devotes some little thought to the apparently insuperable problem of how to reconcile the persistence of individual consciousnesses with their assimilation to Omega. But the problem yields to the application of 'remorseless logic'. The individual particles of consciousness do not join up any old how, but only centre to centre, thanks to the mediation of Love; Omega, then, 'in its ultimate principle, can only be a distinct Centre radiating at the core of a system of centres', and the final state of the world is one in which 'unity coincides with a paroxysm of harmonized complexity'. And so our hero escapes from his dire predicament: with one bound Jack was free.

Although elsewhere Teilhard has dared to write an equation so explicit as 'Evolution = Rise of Consciousness' he does not go so far as to write 'Omega = God'; but in the course of some obscure pious rant he does tell us that God, like Omega, is a 'Centre of centres', and in one place he refers to 'God-Omega'.

How have people come to be taken in by *The Phenomenon of Man*? We must not underestimate the size of the market for works of this kind, for philosophy-fiction. Just as compulsory primary education created a market catered for by cheap dailies and weeklies, so the spread of secondary and latterly of tertiary education has created a large population of people, often with well-developed literary and scholarly tastes, who have been educated far beyond their capacity to undertake analytical thought. It is through their eyes that we must attempt to see the attractions of Teilhard, which I shall jot down in the order in which they come to mind.

1 *The Phenomenon of Man* is anti-scientific in temper (scientists are shown up as shallow folk skating about on the surface of things), and, as if that were not recommendation enough, it was written by a scientist, a fact which seems to give it particular authority and weight. Laymen firmly believe that scientists are one species of person. They are not to know that the different branches of science require very different aptitudes and degrees of skill for their prosecution. Teilhard practised an intellectually unexacting kind of science in which he achieved a moderate proficiency. He has no grasp of what makes a logical argument or of what makes for proof. He does not even preserve the common decencies of scientific writing, though his book is professedly a scientific treatise.

2 It is written in an all but totally unintelligible style, and this is construed as prima-facie evidence of profundity. (At present this applies only to works of French authorship; in later Victorian and Edwardian times the same deference was thought due to

Germans, with equally little reason.) It is because Teilhard has such wonderful *deep* thoughts that he's so difficult to follow— really it's beyond my poor brain but doesn't that just *show* how profound and important it must be?

3 It declares that Man is in a sorry state, the victim of a 'fundamental anguish of being', a 'malady of space-time', a sickness of 'cosmic gravity'. The Predicament of Man is all the rage now that people have sufficient leisure and are sufficiently well fed to contemplate it, and many a tidy literary reputation has been built upon exploiting it; anybody nowadays who dared to suggest that the plight of man might not be wholly desperate would get a sharp rap over the knuckles in any literary weekly. Teilhard not only diagnoses in everyone the fashionable disease but propounds a remedy for it—yet a remedy so obscure and so remote from the possibility of application that it is not likely to deprive any practitioner of a living.

4 *The Phenomenon of Man* was introduced to the English-speaking world by Sir Julian Huxley, who, like myself, finds Teilhard somewhat difficult to follow ('If I understood him aright'; 'here his thought is not fully clear to me'; etc.).[3] Unlike myself, Sir Julian finds Teilhard in possession of a 'rigorous sense of values', one who 'always endeavoured to think concretely'; he was speculative, to be sure, but his speculation was 'always disciplined by logic'. But then it does not seem to me that Huxley expounds Teilhard's argument; his Introduction does little more than to call attention to parallels between Teilhard's thinking and his own. Chief among these is the cosmic significance attached to a suitably generalized conception of evolution—a conception so diluted or attenuated in the course of being generalized as to cover all events or phenomena that are not immobile in time.[4] In particular, Huxley applauds the, in my opinion, mistaken belief that the so-called 'psychosocial evolution' of mankind and the genetical evolution of living organisms generally are two episodes of a continuous integral process (though separated by a 'critical point', whatever that may mean). Yet for all this Huxley finds it impossible to follow Teilhard 'all the way in his gallant attempt to reconcile the supernatural elements in Christianity with the facts

and implications of evolution'. But, bless my soul, this reconciliation is just what Teilhard's book is *about*!

I have read and studied *The Phenomenon of Man* with real distress, even with despair. Instead of wringing our hands over the Human Predicament, we should attend to those parts of it which are wholly remediable, above all to the gullibility which makes it possible for people to be taken in by such a bag of tricks as this. If it were an innocent, passive gullibility it would be excusable; but all too clearly, alas, it is an active willingness to be deceived.

2 Hypothesis and imagination

There is a mask of theory over the whole face of nature.

1

If an educated layman were asked to set down his understanding of what goes on in the head when scientific discoveries are made and of what it is about a scientist that qualifies him to make them, his account of the matter might go something like this. A scientist is a man who has cultivated (if indeed he was not born with) the restless, analytical, problem-seeking, problem-solving temperament that marks his possession of a Scientific Mind. Science is an immensely prosperous and successful enterprise—as religion is not, nor economics (for example), nor philosophy itself—because it is the outcome of applying a certain sure and powerful method of discovery and proof to the investigation of natural phenomena: *The Scientific Method.* The scientific method is not deductive in character—it is a well-known fallacy to regard it as such—but it is rigorous nevertheless, and logically conclusive. Scientific laws are *in*ductive in origin. An episode of scientific discovery begins with the plain and unembroidered evidence of the senses—with innocent, unprejudiced observation, the exercise of which is one of the scientist's most precious and distinctive faculties—and a great mansion of natural law is slowly built upon it. Imagination kept within bounds may ornament a scientist's thought and intuition may bring it faster to its conclusions, but in a strictly formal sense neither is indispensable. Yet Newton was too severe upon hypotheses, for though there is indeed something *mere* about hypotheses, the best of them may look forward to a dignified middle age as Theories.[1]

A critic anxious to find fault might now raise a number of objections, among them these: (1) there is no such thing as a Scientific Mind; (2) there is no such thing as The Scientific Method; (3) the idea of naïve or innocent observation is philosophers' make-believe; (4) 'induction' in the wider sense that Mill gave it is a myth; and (5) the formulation of a natural 'law' always begins as an imaginative exploit, and without imagination scientific thought is barren. Finally (he might add) it is an unhappy usage that treats a hypothesis as an adolescent theory.

1 *There is no such thing as a Scientific Mind.* Scientists are people of very dissimilar temperaments doing different things in very different ways. Among scientists are collectors, classifiers and compulsive tidiers-up; many are detectives by temperament and many are explorers; some are artists and others artisans. There are poet-scientists and philosopher-scientists and even a few mystics. What sort of mind or temperament can all these people be supposed to have in common? *Obligative* scientists must be very rare, and most people who are in fact scientists could easily have been something else instead.

2 *There is no such thing as The Scientific Method*—as *the* scientific method, that is the point: there is no one rounded art or system of rules which stands to its subject-matter as logical syntax stands towards any particular instance of reasoning by deduction. 'An art of discovery is not possible,' wrote a former Master of Trinity; 'we can give no rules for the pursuit of truth which shall be universally and peremptorily applicable.' To many philosophers of science such an opinion must have seemed treasonable, and we can understand their unwillingness to accept a judgement that seems to put them out of business. The face-saving formula is that although there is indeed a Scientific Method, scientists observe its rules unconsciously and do not understand it in the sense of being able to put it clearly into words.

3 *The idea of naïve or innocent observation is philosophers' make-believe.* To good old British empiricists it has always seemed self-

evident that the mind, uncorrupted by past experience, can passively accept the imprint of sensory information from the outside world and work it into complex notions; that the candid acceptance of sense-data is the elementary or generative act in the advancement of learning and the foundation of everything we are truly sure of.[2] Alas, unprejudiced observation is mythical too. In all sensation we pick and choose, interpret, seek and impose order, and devise and test hypotheses about what we witness. Sense data are taken, not merely given: we *learn* to perceive.[3] 'Why can't you draw what you see?' is the immemorial cry of the teacher to the student looking down the microscope for the first time at some quite unfamiliar preparation he is called upon to draw. The teacher has forgotten, and the student himself will soon forget, that what he sees conveys no information until he knows beforehand the kind of thing he is expected to see. I cite more evidence on this point below.

4 *Induction is a myth*. In donnish conversation we are not taken aback when someone says he has 'deduced' something or has carried out a deduction; but if he were to say he had *in*duced something or other we should think him facetious if not a pompous idiot. So it is with 'Laws': scientists do not profess to be trying to discover laws and use the word itself only in conventional contexts (Hooke's Law, Boyle's Law). (The actual usages of scientific speech are, as I shall explain below, extremely revealing.) It is indeed a myth to suppose that scientists actually carry out inductions or that a logical autopsy upon a completed episode of scientific research reveals in it anything that could be called an inductive structure of thought.

'Induction' in the wider sense that distinguishes it from perfect or merely iterative induction (see below) is a word lacking the qualities that would justify its retention in a professional vocabulary. It is seldom, if ever, used in any sentence of which it is not itself the subject, and it has no agreed meaning. *Finding* a meaning for induction has been a philosophic pastime for more than a hundred years. Whewell used the word, but with some feeling in later years that he might have dropped it. 'There is really no such thing as a distinct process of induction,' said Stanley Jevons; 'all inductive reasoning is but the inverse application of deductive reasoning'—and this was what Whewell meant when he said that induction and deduction went upstairs and downstairs on the same staircase. For Samuel Neil, however, 'induction' was confined to the act of testing a scientific conjecture or presupposition, and this was also C. S. Peirce's usage ('The operation of testing a hypothesis by experiment . . . I call induction').[4] Peirce accordingly uses the words *retroduction* or *abduction* to mean what Jevons called *in*duction. Nowadays the tendency is to use 'experimentation' to stand for the acts used in testing a hypothesis, leaving 'induction' as a vague word to signify all the various ways of travelling upstream of the flow of deductive inference. (Popper, of course, is for abandoning 'induction' altogether.)

The word *experiment* has also changed its meaning. When amateurs of the history of science attribute to Bacon the advocacy of the experimental method, they are often acting under the impression that Bacon used the word as we do. But a Baconian 'experiment' had the connotation that still persists in the French 'expérience' today: a Baconian experiment is a contrived experience or contrived happening as opposed to a natural experience or happening, for Bacon rightly supposed that common knowledge was not enough and that there was no relying upon luck of observation—upon 'the casual felicity of particular events'. The Philosophers of Mind took the same view: experiments were 'designed observations' intended 'to place nature in situations in which she never presents herself spontaneously to view, and to extort from her secrets over which she draws a veil to the eyes of others'.[5] Rubbing two sticks together to see what happens is an experiment in Bacon's sense; rubbing two sticks together to see if enough heat can be generated by friction to ignite them is an experiment in the modern sense. An experiment of the first kind leaves one with no answer to the question (a 'good' question: see below), 'Why on earth are you rubbing those two sticks together?'

I shall refer later to the changing connotation of 'hypothesis', a word that has grown in stature as 'induction' has declined.

The concept of induction was entrenched into scientific methodology through the formidable advocacy of John Stuart Mill.

Mill, said John Venn in 1907,[6] 'had 'dominated the thought and study of intelligent students to an extent which many will find it hard to realise at the present day'; yet he could still take a general familiarity with Mill's views for granted, in spite of having recorded as far back as 1889 'a broadening current of dissatisfaction' on the part of physicists which had 'mostly taken the form of an ill-concealed or openly avowed contempt of the logical treatment of Induction'. It is, however, the indifference rather than the hostility of critically-minded scientists that has allowed the myth of induction to persist—combined, I believe, with the great earnestness and sincerity of Mill himself; for Mill believed, as so many good people believe today, that if only we could formulate and master The Scientific Method many of the vexed problems of modern society would vanish before its use.

Mill's was, of course, Induction in the strong, imperfect, or open-ended sense. 'Induction', said Mill ('that great mental operation'), 'is a process of inference; it proceeds from the known to the unknown; and any operation involving no inference, any process in which what seems the conclusion is no wider than the premises from which it is drawn, does not fall within the meaning of the term.' That would be very well if he had not also said that induction was an exact and logically rigorous process, capable of doing for empirical reasoning what logical syntax does for the process of deduction: 'The business of inductive logic is to provide rules and models (such as the syllogism and its rules are for ratiocination) to which, if inductive arguments conform, those arguments are conclusive, and not otherwise.'[7]

There seems no point in mulling over the logical errors of Mill's *System*, for they are now common knowledge—for example, his failure to distinguish between the methodologies of discovery and of proof (though Whewell had insisted on the distinction), and the circularity of his attempt to justify that 'ultimate syllogism' which had 'for its major premise the principle or axiom of the uniformity of the course of nature'. But one may yet be surprised by how little he understood the methodological functions of hypotheses, and by the hopeless ambition embodied in his belief that it was possible merely by taking thought to arrive with certainty at the truth of general statements containing more information than the sum of their known instances.

The current of informed opinion was already flowing in the other direction. The probationary character of scientific law is implicit in all of Whewell, and long before him George Campbell, in his influential and widely read *Philosophy of Rhetoric* (1776), had said of inductive generalization that there 'may be in every step, and commonly is, less certainty than in the preceding; *but in no instance whatever can there be more*' (my italics). 'No hypothesis', said Dugald Stewart, 'can completely exclude the possibility of exceptions or limitations hitherto undiscovered.' By the latter half of the nineteenth century the point had become commonplace. 'No inductive conclusions are more than probable,' said Jevons: 'we never escape the risk of error altogether.' Venn took pains to emphasize his belief 'that no ultimate objective certainty, such as Mill for instance seemed to attribute to the results of induction, is attainable by any exercise of the human reason'. 'The conclusions of science make no pretence to being more than probable,' wrote C. S. Peirce.

The logical status of deduction and syllogistic reasoning had not been seriously in question since the days of Bacon. Syllogistic reasoning (an 'unnatural art', Campbell had called it, and others 'futile' or 'puerile') was indeed a logically conclusive process, but that was because it merely 'expands and unfolds', merely brings to light and makes explicit the information lying more or less deeply hidden in the premises out of which it flows. Deduction makes known to us only what the infirmity of our powers of reasoning has so far left concealed. The case had been well put by Archbishop Whately,[8] and Mill accepted it; and so the peculiar and distinctive role of deduction in scientific reasoning came to be overlooked. Convinced nevertheless that Science had come upon irrefragable general truths by some process other than deduction, Mill had no alternative but to put his faith in induction—to believe in the existence of a valid inductive process even if his own account of it should prove faulty or incomplete.

What about Baconian induction—the painstaking assembly and classification of natural and elicited (experimental) facts of which

Jevons said that it reduced the methodology of science to a kind of bookkeeping? By the sixth edition of the *Origin* in 1876, Darwin had convinced himself that he had been a good Baconian, but his correspondence tells a different story. Darwin's status as the culture-hero of induction—the great but deeply humble scientist listening attentively to Nature's lessons from her own lips—has now to be reconciled with evidence that he had the germ of the idea of natural selection before ever he had read Malthus.[9]

It is Karl Pearson whose scientific practice and theoretical professions earn him the right to be called a true Baconian. 'The classification of facts', he wrote in *The Grammar of Science*, and

the recognition of their sequence and relative significance is the function of science . . . let us be quite sure that whenever we come across a conclusion in a scientific work which does not flow from the classification of facts, or which is not directly stated by the author to be an assumption, then we are dealing with bad science.[10]

Poor Pearson! His punishment was to have practised what he preached, and his general theory of heredity, of genuinely inductive origin, was in principle quite erroneous.

I have given here the conventional view of Bacon's methodology, and shall return later to the claim made on his behalf by Coleridge and others that he was fully aware of the methodological value of hypotheses.

5 *The formulation of a natural law begins as an imaginative exploit and imagination is a faculty essential to the scientist's task.* Most words of the philosopher's vocabulary, including 'philosopher' itself, have changed their usages over the past few hundred years.[11] 'Hypothesis' is no exception. In a modern professional vocabulary a hypothesis is an imaginative preconception of *what might be true* in the form of a declaration with verifiable deductive consequences. It no longer tows 'gratuitous', 'mere', or 'wild' behind it, and the pejorative usage ('Evolution is a mere hypothesis', 'It is only a hypothesis that smoking causes lung cancer') is one of the outward signs of little learning. But in the days of Travellers' Tales and Marvels, when (as John Gregory contemptuously remarked)[12] philosophers were more interested in animals with two heads than in animals with one, 'hypothesis' carried very strongly

the connotation of the wantonly fanciful and above all the gratuitous; nor was there any thought that a hypothesis need do more
than explain the phenomena it was expressly formulated to explain. The element of *responsibility* that goes with the formulation
of a hypothesis today was altogether lacking. Thomas Burnet's
Sacred Theory of the Earth (1684–90) is a case in point—a romantic
and absurd cosmology using the word 'hypothesis' in just the
sense that Newton repudiated. 'Men of short thoughts and little
meditation,' Burnet says in self-defence, 'call such theories as
these, Philosophick Romances.' But, he says:

there is no surer mark of a good Hypothesis, than when it doth not only
hit luckily in one or two particulars but answers all that it is to be applied
to, and is adequate to Nature in her whole extent.
   But how fully or easily soever these things may answer Nature, you will
say, it may be, that all this is but an Hypothesis; that is, a kind of fiction
or supposition that things were so and so at first, and by the coherence
and agreement of the Effects with such a supposition, you would argue
and prove that this is so. This I confess is true, this is the method, and if
we would know anything in Nature further than our senses go, we can
know it no otherwise than by an Hypothesis . . . and if that Hypothesis be
easie and intelligible, and answers all the phaenomena . . . you have done
as much as a Philosopher or as Humane reason can do.

Burnet's reasoning thus ends at the very point at which scientific reasoning begins. He did not seem to realize that his hypothesis about what the Earth was like before the flood and what it
would be like after the Fire was but one among a virtual infinitude
of hypotheses, and that he was under a moral obligation to find
out if his were preferable to any other.

Burnet's preposterous speculations were expounded in prose
that earned him immortality; because they were not, many philosophic romances that must have been known to Newton are now
forgotten. Thomas Reid shall be allowed to sum the prevailing
situation up.[13] 'It is genius,' he says, 'and not the want of it, that
adulterates philosophy, and fills it with error and false theory. A
creative imagination disdains the mean offices of digging for a
foundation,' leaving these servile employments to scientific
drudges. 'The world has been so long befooled by hypotheses in
all parts of philosophy,' that we must learn 'to treat them with just

contempt, as the reveries of vain and fanciful men.' Newton *could* have invented a hypothesis to account for gravitation, but 'his philosophy was of another complexion', for Newton had been 'taught by Lord Bacon to despise hypotheses as fictions of human fancy'.

Because Newton is cast as the hero of every scientific methodology of the past two hundred years, philosophers who attached great importance to hypotheses felt it their duty to explain away Newton's famous and profoundly influential disavowal. Stanley Jevons was so sure that Newton had practised what is now often called the 'hypothetico-deductive' method that he was inclined to think *hypotheses non fingo* ironical. But for two hundred years after Newton no one could advocate the use of hypotheses without an uneasy backward glance. Dugald Stewart said that an 'indiscriminate zeal against hypotheses' had been 'much encouraged by the strong and decided terms in which, on various occasions, they are reprobated by Newton'. 'Newton appears to have had a horrour of the term *hypothesis*,' said William Whewell. Sir John Herschel spoke up in favour of hypotheses.[14] Samuel Neil in 1851 deplored the 'widely prevalent prejudice in the present age against hypoheses', and Thomas Henry Huxley had felt obliged to say, 'Do not allow yourselves to be misled by the common notion that a hypothesis is untrustworthy merely because it is a hypothesis.' Even George Henry Lewes found himself unable to propound his fairly sensible views on hypotheses without much prevarication and pursing of the lips.[15]

Where does Mill stand? Modern philosophers who are for various reasons 'pro-Mill' can of course find him a devotee of hypotheses. Hypotheses, Mill will be found to say, provide 'temporary aid', even 'large temporary assistance' ('temporary' because hypotheses are the larval forms of theories); hypotheses are valuable because they suggest observations and experiments, and in this respect they are indeed indispensable. However, all this had been said before, repeatedly: some instances I shall cite later. In his less conventional utterances on hypotheses Mill betrayed that he had no deep understanding of what is now thought to be their distinctive methodological function. He feared that people who used hypotheses did so under the impression that a hypothesis must be true if the inferences drawn from it were in accordance with the facts. Later therefore he says: 'It seems to be thought that an hypothesis . . . is entitled to a more favourable reception, if besides accounting for all the facts previously known, it has led to the anticipation and prediction of others which experience afterwards verified.'[16] This, he says, is 'well calculated to impress the uninformed', but will not impress thinkers 'of any degree of sobriety'.

Mill feared the imaginative element in hypotheses: 'a hypothesis being a mere supposition, there are no other limits to hypotheses than those of the human imagination'. These are Reid's fears, preying on Mill at a time when good reasons for feeling fearful had largely disappeared. Today we think the imaginative element in science one of its chief glories. Even Karl Pearson recognized it as a motive force in *great* discoveries, but, of course, 'imagination must not replace reason in the deduction of relation and law from classified facts'. (The belief that great discoveries and little everyday discoveries have quite different methodological origins betrays the amateur. Whewell, the professional, insisted that the bold use of the imagination was the rule in scientific discovery, not the exception: see below.) All the same, the idea that hypotheses arose by mere conjecture, by guesswork, was thought undignified. Whewell had called good hypotheses 'happy guesses', though elsewhere, as if the occasion called for something more formal, he spoke of 'felicitous strokes of inventive talent'. But philosophers like Venn did not take to it: 'it is . . . scarcely an exaggeration of Whewell's account of the inductive process to say of it, as in fact has been said, that it simply resolves itself into making guesses'.

It is the word that is at fault, not the conception. To say that Einstein formulated a theory of relativity by guesswork is on all fours with saying that Wordsworth wrote rhymes and Mozart tuneful music. It is cheeky where something grave is called for.

2

I now turn to consider the history during the eighteenth and nineteenth centuries of some of the central ideas of the hypothetico-deductive scheme of scientific reasoning, confining myself, as hitherto, almost wholly to English and Scottish philosophers and the tradition of thought they embody. Among these ideas are:

(1) the uncertainty of all 'inductive' reasoning and the probationary status of hypotheses;

(2) the role of the hypothesis in starting enquiry and giving it direction, so confining the domain of observation to something smaller than the whole universe of observables;

(3) the asymmetry of proof and disproof: only disproof is logically conclusive;

(4) the obligation to put a hypothesis to the test.

1 I have already mentioned a number of earlier opinions on the inconclusiveness of scientific reasoning (above, p. 17).

2 It is our imaginative preconception of what might be true that gives us an incentive to seek the truth and a clue to where we might find it. 'In every useful experiment,' said John Gregory, writing in 1772, 'there must be some point in view, some anticipation of a principle to be established or rejected.' Such anticipations, he went on to say, are *hypotheses*: people were suspicious of hypotheses because they did not fully understand their purpose, but without them 'there could not be useful observation, nor experiment, nor arrangement, because there would be no motive or principle in the mind to form them'. Dugald Stewart quoted passages expressing the same opinion in the writings of Boscovich, Robert Hooke, and Stephen Hales[17]—scientists all three. But on this point Coleridge sweeps everyone else aside.[18] In every advance of science, he assures us, 'a previous act and conception of the mind . . . an *initiative* is indispensably necessary', for when it comes to founding a theory on generalization, 'what shall determine the mind to one point rather than another; within what limits, and from what number of individuals shall the generalization be made? *The theory must still require a prior theory for*

its own legitimate construction' (my italics). Coleridge (like Stewart and later Neil) managed to convince himself of the great Francis Bacon's full awareness of the need for an 'intellectual or mental *initiative'* as the 'motive and guide of every philosophical experiment . . . namely, some well-grounded purpose, some distinct impression of the probable results, some self-consistent anticipation . . . which he affirms to be the prior *half* of the knowledge sought, *dimidium scientiae'*. The passage all three quote[19] as evidence for this interpretation is, in my reading of it, too slight to carry so great a weight of meaning.

3 Many philosophers in the older and the newer senses have spoken of the value of false hypotheses, and Stewart particularly commends the opinions of Boscovich ('the slightest of whose logical hints are entitled to particular attention'). Boscovich had said that by means of hypotheses

> we are enabled to supply the defects of our *data*, and to conjecture or divine the path to truth; always ready to abandon our hypothesis, when found to involve consequences inconsistent with fact. And, indeed, in most cases, I conceive this to be the method best adapted to physics; a science in which . . . legitimate theories are generally the slow result of disappointed essays, and of errors which have led the way to their own detection.[20]

This is all right as far as it goes, but what one will not find so easily is a premonition of one of the strongest ideas in Popper's methodology, that the only act which the scientist can perform with complete logical certainty is the repudiation of what is false. It is *falsification* that has the logical stature attributed by the logical positivists to verification. 'Every experiment may be said to exist only in order to give the facts a chance of disproving the null hypothesis.'[21] The asymmetry of proof, considered as a point of logic, is of course very elementary, and it is merely slovenly or simple-minded to suppose that hypotheses are proved true if they lead to true conclusions. No logician of science has ever done so. Whewell certainly realized that refutation was methodologically a strong procedure, stronger than confirmation, an opinion that comes out more clearly in his aphorisms than in the body of the text:

(ix) The truth of tentative hypotheses must be tested by their application to facts. The discoverer must be ready, carefully to try his hypotheses in this manner . . . and to reject them if they will not bear the test.

(x) The process of scientific discovery is cautious and rigorous, not by abstaining from hypotheses, but by rigorously comparing hypotheses with facts, and by resolutely rejecting all which the comparison does not confirm.

These opinions shocked Mill. Dr Whewell's system, he complained, did not recognize 'any necessity for proof': 'If, after assuming an hypothesis and carefully collating it with facts, nothing is brought to light inconsistent with it, that is, if experience does not *dis*prove it, he is content; at least until a simpler hypothesis, equally consistent with experience, presents itself.' To Mill this attitude of Whewell's betrayed 'a radical misconception of the nature of the evidence of physical truths'. No wonder Venn said that a person who read both Mill and Whewell would find it hard to believe that they were discussing the same subject! The verdict must go to Whewell, 'whose acquaintance with the processes of thought of science', said Peirce, 'was incomparably greater than Mill's'.

4 The formulation of a hypothesis carries with it an obligation to test it as rigorously as we can command skills to do so. There was no sign of any such sense of obligation in Burnet's *Sacred Theory*: to explain the phenomena it was designed to explain was judged evidence enough. It satisfied curiosity in much the same way as a mother's desperately *ad hoc* answers satisfy the insistent questioning of a child. The child is not interested in the content of the answer: he asks as if he were under an instinctual compulsion to do so, and the act of answering completes a sort of ritual of exploration. But when curiosity is satisfied it is discharged: formulation of a hypothesis may act as a deterrent rather than as a stimulus to enquiry—a danger the earlier critics of the use of hypotheses were fully aware of.

Even the more sophisticated authors of Philosophick Romances did not seem to realize that any one set of phenomena could be explained by many hypotheses other than the one they fancied. It seems a strange blindness, but I think that Dugald Stewart in a finely reasoned passage got to the bottom of it. It was a favourite conceit in eighteenth-century philosophizing—Stewart found it in Boscovich, Le Sage, D'Alembert, Gravesande and Hartley—that natural philosophy is, in David Hartley's words, 'the art of *decyphering* the Mysteries of Nature . . . so that . . . every Theory which can explain all the Phaenomena, has all the same Evidence in its favour, that it is possible the Key of a Cypher can have from its explaining that Cypher'.[22] Stewart found the analogy inept for many reasons, the chief being that whereas a cypher has one key, a unique solution, physical hypotheses seldom, if ever, 'afford the *only* way of explaining the phenomena to which they are applied'.

It is all very well to say that we are under a permanent obligation to test hypotheses, but, as Peirce said, 'there are some hypotheses which are of such a nature that they can never be tested at all. Whether such hypotheses ought to be entertained at all, and if so in what sense, is a serious question.' Certainly the logical positivists took the question very seriously indeed, and Popper has done so too, but I do not recollect its having been a live issue before Peirce.

3

Let me now set out the gist of the hypothetico-deductive system as it might be formulated today. ('Gist' is the right word, for there is no question of its providing an abstract formal framework which becomes a concrete example of scientific reasoning when we fill in the blanks.) First, there is a clear distinction between the acts of mind involved in discovery and in proof. The generative or elementary act in discovery is 'having an idea' or proposing a hypothesis. Although one can put oneself in the right frame of mind for having ideas and can abet the process, the process itself is outside logic and cannot be made the subject of logical rules. Hypotheses must be tested, that is criticized. These tests take the form of finding out whether or not the deductive consequences of the hypothesis or systems of hypotheses are statements that correspond to reality. As the very least we expect of a hypothesis is that it should account for the phenomena already before us, its 'extra-mural' implications, its predictions about what is not yet known to be the case, are of special and perhaps crucial importance. If the predictions are false, the hypothesis is wrong or in need of modification; if they are true, we gain confidence in it, and can, so to speak, enter it for a higher examination; but if it is of such a kind that it cannot be falsified even in principle, then the hypothesis belongs to some realm of discourse other than science. Certainty can be aspired to, but a 'rightness' that lies beyond the possibility of future criticism cannot be achieved by any scientific theory. There is no place for apodictic certainty in science.

The first strongly reasoned and fully argued exposition of a hypothetico-deductive system is unquestionably Karl Popper's. Quite a large part of it had been propounded at the level of learned discourse rather than of critical analysis by William Whewell, FRS, Master of Trinity College, Cambridge, in 1840. Whewell is never heard of nowadays outside the ranks of historians of science: if one mentions his name one may be asked to spell it. But his reputation in his day was formidable. Whewell wrote upon ethics, hydrostatics, political economy, astronomy, verse composition, terminology, the Platonic dialogues, mechanics, geology, and the History and Philosophy of the Inductive Sciences. He was the first *scientist*, I believe, to express a lengthy and carefully thought out opinion on the nature of scientific discovery, and in a sense the first scientist of any description, for he invented the word itself.

There are many inadequacies in Whewell, but the spirit is right. No general statement, he said, not even the simplest iterative generalization, can arise merely from the conjunction of raw data. The mind always makes some imaginative contribution of its own, always 'superinduces' some idea upon the bare facts. A hypothesis is an explanatory conjecture giving one of many possible explanations that might meet the case.

A facility in devising hypotheses, therefore, is so far from being a fault in the intellectual character of a discoverer, that it is, in truth, a faculty indispensable to his task.

To form hypotheses, and then to employ much labour and skill in refuting, if they do not succeed in establishing them, is a part of the usual process of inventive minds. Such a proceeding belongs to the rule of the genius of discovery, rather than (as has often been taught in modern times) to the *exception*.

Yet it is indispensably necessary for the discoverer to demand of his hypotheses 'an agreement with facts such as will withstand the most patient and rigid inquiry', and, if they are found wanting, to turn them resolutely down:

Since the discoverer has thus constantly to work his way onwards by means of hypotheses, false and true, it is highly important for him to possess talents and means for rapidly *testing* each supposition as it offers itself.

The hypotheses which we accept ought to explain phenomena which we have observed. But they ought to do more than this: our hypotheses ought to *foretell* phenomena which have not yet been observed [but which are] of the same kind as those which the hypothesis was invented to explain.

Whewell did not believe that a scientist acquired factual information by passive attention to the evidence of his senses; the idea of 'naïve' or 'innocent' observation (see above, pp. 13–14) he rejected altogether: 'Facts cannot be observed as Facts except in virtue of the Conceptions which the observer himself unconsciously supplies.' The distinction between fact and theory was by no means as distinct as people were accustomed to believe: 'There is a mask of theory over the whole face of nature.' Strictly speaking, no scientific discovery can be made by accident. What Whewell has to say on Man as the Interpreter of Nature[23] is a suitable prolegomenon to Popper's famous lecture 'On the Sources of Knowledge and of Ignorance'.

The account of scientific method which became recognized as the official alternative and rival to Mill's was not Whewell's but Stanley Jevons's. Jevons is not as fresh as Whewell nor so boldly original; we may think he should have acknowledged Whewell more often than he did. Jevons gave it as his 'very deliberate opinion' that 'many of Mill's innovations in logical science . . . are entirely groundless and false'. As to Bacon, he took the 'extreme view of holding that Francis Bacon . . . had no correct notions as to the logical method by which from particular facts we deduce laws of nature'. Jevons endeavoured to show that 'hypothetical anticipation of nature is an essential part of inductive inquiry', the method 'which has led to all the great triumphs of scientific research'. Even in the 'apparently passive observation of a phenomenon' our attention should be 'guided by theoretical anticipations'.

The three essential stages in the process which he continued with deliberate vagueness to call 'induction' were, using his own words,

(*a*) Framing some hypothesis as to the character of the general law.

(*b*) Deducing consequences from that law.

(*c*) Observing whether the consequences agree with the particular facts under consideration.

Hypothesis is always employed, he says, consciously or unconsciously.

This account of the matter had come to be pretty widely agreed upon during the second half of the nineteenth century. We shall find it in Neil and Adamson[24] and very clearly in Peirce. (Venn, in spite of his reputation, I find disappointing.) There are many premonitions of the hypothetico-deductive method in the eighteenth century and even earlier, particularly in the writing of scientists. The clearest known to me is Dugald Stewart's, a point worth making because of the dismissive and totally erroneous opinion that his philosophy is simply a reproduction of his master's, Thomas Reid's, voice. In answer to Reid's rhetorical challenge to name any advance in science which had arisen by the use of a hypothetical method, Stewart thought it sufficient to mention the theory of gravitation and the Copernican system.

Stewart believed that most discoveries in science had grown out of hypothetical reasoning:

It is by reasoning synthetically from the hypothesis, and comparing the deductions with observation and experiment, that the cautious inquirer is gradually led, either to correct it in such a manner as to reconcile it with facts, or finally to abandon it as an unfounded conjecture. Even in this latter case, an approach is made to the truth in the way of *exclusion* . . .

Stewart's own analysis of the use of those tiresome adjectives *synthetic* and *analytic* shows he is here using 'synthetically' in the sense of 'deductively'.

4

A scientific methodology, being itself a theory about the conduct of scientific enquiry, must have grown out of an attempt to find out exactly what scientists do or ought to do. The methodology should therefore be measured against scientific practice to give us confidence in its worth. Unfortunately, this honest ambition is fraught with logical perils. If we assume for the sake of argument that the methodology is unsound, then so also will be our test of its validity. If we assume it to be sound, then there is no point in submitting it to test, for the test could not invalidate it. These difficulties I shall surmount by disregarding them entirely.

What scientists *do* has never been the subject of a scientific, that is, an ethological enquiry. It is no use looking to scientific 'papers', for they not merely conceal but actively misrepresent the reasoning that goes into the work they describe.[25] If scientific papers are to be accepted for publication, they must be written in the inductive style. The spirit of John Stuart Mill glares out of the eyes of every editor of a Learned Journal.

Nor is it much use listening to accounts of what scientists *say* they do, for their opinions vary widely enough to accommodate almost any methodological hypothesis we may care to devise. Only unstudied evidence will do—and that means listening at a keyhole. Here are some turns of speech we may hear in a biological laboratory:

'What gave you the idea of trying . . . ?'
'I'm taking the view that the underlying mechanism is . . .'
'What happens if you assume that . . . ?'
'Actually, your results can be accounted for on a quite different hypothesis.'
'It follows from what you are saying that if . . . , then . . .'
'Is that actually the case?'

'That's a good question.' [i.e. a question about a true weakness, insufficiency or ambiguity]

'That result squared with my hypothesis.'

'So obviously that idea was out.'

'At the moment I don't see any way of eliminating that possibility.'

'My results don't make a story yet.'

'I'm still at the stage of trying to find out if there is anything to be explained.'

'Obviously a great deal more work has got to be done before . . .'

'I don't seem to be getting anywhere.'

Scientific thought has already reached a pretty sophisticated professional level before it finds expression in language such as this. This is not the language of induction. It does not suggest that scientists are hunting for facts, still less that they are busy formulating 'laws'. Scientists are building explanatory structures, *telling stories* which are scrupulously tested to see if they are stories about real life.

It has been a tradition among philosophers that we should look to the physical sciences and to simple, lofty discoveries if we are to see the Scientific Method at work in its most easily intelligible form. I question this opinion. The simplicity of great discoveries is often a measure of how far they have travelled from their beginnings. Let a biologist have a turn. Here is Claude Bernard, writing just one hundred years ago: 'A hypothesis is . . . the obligatory starting point of all experimental reasoning. Without it no investigation would be possible, and one would learn nothing: one could only pile up barren observations. To experiment without a preconceived idea is to wander aimlessly.' Indeed, 'Those who have condemned the use of hypotheses and preconceived ideas in the experimental method have made the mistake of confusing the contriving of the experiment with the verification of its results.'[26] Over and over again Bernard insists that hypotheses must be of such a kind that they can be tested, that one should go out of one's way to find means of refuting them, and that 'if one proposes a hypothesis which experience cannot verify, one abandons the experimental method'. Claude Bernard is most distinctive and at his best in his insistence on the critical method, on the virtue and necessity of Doubt.

When propounding a general theory in science, the one thing one can be sure of is that, in the strict sense, such theories are mistaken. They are only partial and provisional truths which are necessary . . . to carry the investigation forward; they represent only the current state of our understanding and are bound to be modified by the growth of science . . .

This is powerful evidence, for Claude Bernard, in creating experimental physiology, did indeed put scientific medicine on a new foundation. His philosophy *worked*.

In real life the imaginative and critical acts that unite to form the hypothetico-deductive method alternate so rapidly, at least in the earlier stages of constructing a theory, that they are not spelled out in thought. The 'process of invention, trial, and acceptance or rejection of the hypothesis goes on so rapidly,' said Whewell, 'that we cannot trace it in its successive steps'. What then is the point of asking ourselves where the initiative comes from, the observation or the idea? Is it not as pointless as asking which came first, the chicken or the egg?

But this is not a pointless question: it matters terribly which came first: scientific dynasties have been overthrown by giving the wrong answer! It matters no less in methodology: we may collect and classify facts, we may marvel at curiosities and idly wonder what accounts for them, but the activity that is characteristically scientific begins with an explanatory conjecture which at once becomes the subject of an energetic critical analysis. It is an instance of a far more general stratagem that underlies every enlargement of general understanding and every new solution of the problem of finding our way about the world. The regulation and control of hypotheses is more usefully described as a *cybernetic* than as a logical process: the adjustment and reformulation of hypotheses through an examination of their deductive consequences is simply another setting for the ubiquitous phenomenon of negative feedback. The purely logical element in scientific discovery is a comparatively small one, and the idea of a *logic* of scientific discovery is acceptable only in an older and wider use of 'logic' than is current among formal logicians today.

The weakness of the hypothetico-deductive system, in so far as it might profess to offer a complete account of the scientific process, lies in its disclaiming any power to explain how hypotheses come into being. By 'inspiration', surely: by the 'spontaneous conjectures of instinctive reasoning', said Peirce: but what then? It has often been suggested that the act of creation is the same in the arts as it is in science:[27] certainly 'having an idea'—the formulation of a hypothesis—resembles other forms of inspirational activity in the circumstances that favour it, the suddenness with which it comes about, the wholeness of the conception it embodies, and the fact that the mental events which lead up to it happen below the surface of the mind. But there, to my mind, the resemblance ends. No one questions the inspirational character of musical or poetic invention because the delight and exaltation that go with it somehow communicate themselves to others. Something *travels*: we're carried away. But science is not an art form in this sense; scientific discovery is a private event, and the delight that accompanies it, or the despair of finding it illusory, does not travel. One scientist may get great satisfaction from another's work and admire it deeply; it may give him great intellectual pleasure; but it gives him no sense of participation in the discovery, it does not carry him away, and his appreciation of it does not depend on his being carried away. If it were otherwise the inspirational origin of scientific discovery would never have been in doubt.

3 Is the scientific paper a fraud?

I have chosen for my title a question: Is the scientific paper a fraud? I ought to explain that a scientific 'paper' is a printed communication to a learned journal, and scientists make their work known almost wholly through papers and not through books, so papers are very important in scientific communication. As to what I mean by asking 'is the scientific paper a fraud?'—I do not of course mean 'does the scientific paper misrepresent facts', and I do not mean that the interpretations you find in a scientific paper are wrong or deliberately mistaken. I mean the scientific paper may be a fraud because it misrepresents the processes of thought that accompanied or gave rise to the work that is de-scribed in the paper. That is the question, and I will say right away that my answer to it is 'yes'. The scientific paper in its orthodox form does embody a totally mistaken conception, even a travesty, of the nature of scientific thought.

Just consider for a moment the traditional form of a scientific paper (incidentally, it is a form which editors themselves often insist upon). The structure of a scientific paper in the biological sciences is something like this. First, there is a section called the 'introduction' in which you merely describe the general field in which your scientific talents are going to be exercised, followed by a section called 'previous work' in which you concede, more or less graciously, that others have dimly groped towards the funda-mental truths that you are now about to expound. Then a section on 'methods'—that is OK. Then comes the section called 'results'. The section called 'results' consists of a stream of factual infor-mation in which it is considered extremely bad form to discuss the significance of the results you are getting. You have to pretend that your mind is, so to speak, a virgin receptacle, an empty vessel, for information which floods into it from the external world for no reason which you yourself have revealed. You reserve all appraisal of the scientific evidence until the 'discussion' section, and in the discussion you adopt the ludicrous pretence of asking yourself if the information you have collected actually means anything; of asking yourself if any general truths are going to emerge from the contemplation of all the evidence you brandished in the section called 'results'.

Of course, what I am saying is rather an exaggeration, but there is more than a mere element of truth in it. The conception underlying this style of scientific writing is that scientific discovery is an inductive process. What induction implies in its cruder form is roughly speaking this: scientific discovery, or the formulation of scientific theory, starts with the unvarnished and unembroidered evidence of the senses. It starts with simple observation—simple, unbiased, unprejudiced, naïve, or innocent observation—and out of this sensory evidence, embodied in the form of simple propositions or declarations of fact, generalizations will grow up and take shape, almost as if some process of crystallization or condensation were taking place. Out of a disorderly array of facts, an orderly theory, an orderly general statement, will somehow emerge. This conception of scientific discovery in which the initiative comes from the unembroidered evidence of the senses was mainly the work of a great and wise, but in this context, I think, very mistaken man—John Stuart Mill.

John Stuart Mill saw, as of course a great many others had seen before him, including Bacon, that deduction in itself is quite powerless as a method of scientific discovery—and for this simple reason: that the process of deduction as such only uncovers, brings out into the open, makes explicit, information that is already present in the axioms or premises from which the process of deduction started. The process of deduction reveals nothing to us except what the infirmity of our own minds has so far concealed from us. It was Mill's belief that induction was the method of science—'that great mental operation', he called it, 'the operation of discovering and proving general propositions'. And round this conception there grew up an inductive logic, of which the business was 'to provide rules to which, if inductive arguments conform, those arguments are conclusive'. Now John Stuart Mill's deeper motive in working out what he conceived to be the essential method of science was to apply that method to the solution of sociological problems: he wanted to apply to sociology the methods which the practice of science had shown to be immensely powerful and exact.

It is ironical that the application to sociology of the inductive method, more or less in the form in which Mill himself conceived it, should have been an almost entirely fruitless one. The simplest application of the Millsian process of induction to sociology came in a rather strange movement called Mass Observation. The belief underlying Mass Observation was apparently this: that if one could only record and set down the actual raw facts about what people do and what people say in pubs, in trains, when they make love to each other, when they are playing games, and so on, then somehow, from this wealth of information, a great generalization would inevitably emerge. Well, in point of fact, nothing important emerged from this approach, unless somebody has been holding out on me. I believe the pioneers of Mass Observation were ornithologists. Certainly they were man-watching—were applying to sociology the very methods which had done so much to bring ornithology into disrepute.

The theory underlying the inductive method cannot be sustained. Let me give three good reasons why not. In the first place, the starting point of induction, naïve observation, innocent observation, is a mere philosophic fiction. There is no such thing as unprejudiced observation. Every act of observation we make is biased. What we see or otherwise sense is a function of what we have seen or sensed in the past.

The second point is this. Scientific discovery or the formulation of the scientific idea on the one hand, and demonstration or proof on the other hand, are two entirely different notions, and Mill confused them. Mill said that induction was the 'operation of discovering and proving general propositions', as if one act of mind would do for both. Now discovery and proof could depend on the same act of mind, and in deduction they do. When we indulge in the process of deduction—as in deducing a theorem from Euclidian axioms or postulates—the theorem contains the discovery (or, more exactly, the uncovery of something which was there in the axioms and postulates, though it was not actually evident) and the process of deduction itself, if it has been carried out correctly, is also the proof that the 'discovery' is valid, is logically correct. So in the process of deduction, discovery and proof can depend on the same process. But in scientific activity they are not the same thing—they are, in fact, totally separate acts of mind.

But the most fundamental objection is this. It simply is not logically possible to arrive with certainty at any generalization containing more information that the sum of the particular statements upon which that generalization was founded, out of which it was woven. How could a mere act of mind lead to the discovery of new information? It would violate a law as fundamental as the law of conservation of matter: it would violate the law of conservation of information.

In view of all these objections, it is hardly surprising that Bertrand Russell in a famous footnote that occurs in his *Principles of Mathematics* of 1903 should have said that, so far as he could see, induction was a mere method of making plausible guesses. And our greatest modern authority on the nature of scientific method, Professor Karl Popper, has no use for induction at all: he regards the inductive process of thought as a myth. 'There is no need even to mention induction,' he says in his great treatise on *The Logic of Scientific Discovery*—though of course he does.

Now let me go back to the scientific papers. What is wrong with the traditional form of scientific paper is simply this: that all scientific work of an experimental or exploratory character starts with some expectation about the outcome of the enquiry. This expectation one starts with, this hypothesis one formulates, provides the initiative and incentive for the enquiry and governs its actual form. It is in the light of this expectation that some observations are held relevant and others not; that some methods are chosen, others discarded; that some experiments are done rather than others. It is only in the light of this prior expectation that the activities the scientist reports in his scientific papers really have any meaning at all.

Hypotheses arise by guesswork. That is to put it in its crudest form. I should say rather that they arise by inspiration; but in any event they arise by processes that form part of the subject-matter of psychology and certainly not of logic, for there is no logically rigorous method for devising hypotheses. It is a vulgar error, often committed, to speak of 'deducing' hypotheses. Indeed one does not deduce hypotheses: hypotheses are what one deduces things from. So the actual formulation of a hypothesis is—let us say a guess; is inspirational in character. But hypotheses can be tested rigorously—they are tested by experiment, using the word 'experiment' in a rather general sense to mean an act performed to test a hypothesis, that is, to test the deductive consequences of a hypothesis. If one formulates a hypothesis, one can deduce from it certain consequences which are predictions or declarations about what will, or will not, be the case. If these predictions and declarations are mistaken, then the hypothesis must be discarded, or at least modified. If, on the other hand, the predictions turn out correct, then the hypothesis has stood up to trial, and remains on probation as before. This formulation illustrates very well, I think, the distinction between on the one hand the discovery or formulation of a scientific idea or generalization, which is to a greater or lesser degree an imaginative or inspirational act, and on the other hand the proof, or rather the testing of a hypothesis, which is indeed a strictly logical and rigorous process, based upon deductive arguments.

This alternative interpretation of the nature of the scientific process, of the nature of scientific method, is sometimes called the hypothetico-deductive interpretation and this is the view which Professor Karl Popper in *The Logic of Scientific Discovery* has persuaded us is the correct one. To give credit where credit is surely due, it is proper to say that the first professional scientist to express a fully reasoned opinion upon the way scientists actually think when they come upon their scientific discoveries—namely

William Whewell, a geologist, and incidentally the Master of Trinity College, Cambridge—was also the first person to formulate this hypothetico-deductive interpretation of scientific activity. Whewell, like his contemporary Mill, wrote at great length— unnecessarily great length, one is nowadays inclined to think— and I cannot recapitulate his argument, but one or two quotations will make the gist of his thought clear. He said: 'An art of discovery is not possible. We can give no rules for the pursuit of truth which should be universally and peremptorily applicable.' And of hypotheses, he said, with great daring—why it was daring I will explain in just a second—'a facility in devising hypotheses, so far from being a fault in the intellectual character of a discoverer, is a faculty indispensable to his task'. I said this was daring because the word 'hypothesis' and the conception it stood for was still in Whewell's day a rather discreditable one. Hypotheses had a flavour about them of what was wanton and irresponsible. The great Newton, you remember, had frowned upon hypotheses. 'Hypotheses non fingo', he said, and there is another version in which he says 'hypotheses non sequor'—I do not pursue hypotheses.

So to go back once again to the scientific paper: the scientific paper is a fraud in the sense that it does give a totally misleading narrative of the processes of thought that go into the making of scientific discoveries. The inductive format of the scientific paper should be discarded. The discussion which in the traditional scientific paper goes last should surely come at the beginning. The scientific facts and scientific acts should follow the discussion, and scientists should not be ashamed to admit, as many of them apparently *are* ashamed to admit, that hypotheses appear in their minds along uncharted byways of thought; that they are imaginative and inspirational in character; that they are indeed adventures of the mind. What, after all, is the good of scientists reproaching others for their neglect of, or indifference to, the scientific style of thinking they set such great store by, if their own writings show that they themselves have no clear understanding of it?

Anyhow, I am practising what I preach. What I have said about the nature of scientific discovery you can regard as being itself a hypothesis, and the hypothesis comes where I think it should be, namely, it comes at the beginning of the series. Later speakers will provide the facts which will enable you to test and appraise this hypothesis, and I think you will find—I hope you will find—that the evidence they will produce about the nature of scientific discovery will bear me out.

4 The Act of Creation

The author of *Darkness at Noon* must be listened to attentively, no matter what he may choose to write upon. Arthur Koestler is a very clever, knowledgeable, and inventive man, and *The Act of Creation*[1] is very clever too, and full of information, and quite wonderfully inventive in the use of words. Many of the points it makes are not likely to be challenged. That wit and creative thought have much in common; that great syntheses may be come upon by logically unmapped pathways; that putting two and two together is an important element in discovery and also, in a certain sense, in making jokes: it has all been said before, of course, and in fewer than 750 pages, though never with such vitality; and anyhow much of it will bear repeating. But as a serious and original work of learning I am sorry to say that, in my opinion, *The Act of Creation* simply won't do. This is not because of its amateurishness, which is more often than not endearing, nor even because of its blunders—they don't affect Koestler's arguments very much one way or another, even when they reveal a deep-seated misunderstanding of, for example, 'Neo-Darwinism', or find expression in fatuous epigrams like 'All automatic functions of the body are patterned by rhythmic pulsations.' I shall try to explain later what I think wrong with Koestler's technical arguments, but let us first of all examine *The Act of Creation* at the level of philosophical *belles lettres*.

As to style, Koestler overdoes it. On one half-page catharsis is described as an 'earthing' of the emotions, the satisfaction of seeing a joke is said to supply 'added voltage to the original charge detonated in laughter', and a smutty story is put at 'the infra-red end of the emotive spectrum'. We aren't quite sure when he intends to be taken literally: for example, what about 'A concept has as many dimensions in semantic space as there are matrices of which it is a member'? This could have been intended to express an exact idea, for 'space' and 'dimension' have generalized technical meanings; but the feeble passage that follows, illustrated by the word 'Madrid', tells us only that the word itself and the city it stands for conjure up all kinds of different associations in his mind.

When I started Koestler's book I hoped he was going to do something for which his knowledge and sympathies and writer's insight should give him unequalled qualifications: that he would give us a first ethology of scientific activity, and so help to make the scientist intelligible to others and to himself. Unhappily, there are passages in *The Act of Creation* which convince me that he has no real grasp of how scientists go about their work. Consider the depths of misunderstanding revealed by Koestler's aloof and snobbish remarks about the 'unseemly haste' with which some scientists publish their discoveries. Koestler's historical hobnobbings with men of genius seem to have made him forget the fact that, in science, what X misses today Y will surely hit upon tomorrow (or maybe the day after tomorrow). Much of a scientist's pride and sense of accomplishment turns therefore upon being the *first* to do something—upon being the man who did actually speed up or redirect the flow of thought and the growth of understanding. There is no spiritual copyright in scientific discoveries, unless they should happen to be quite mistaken. Only in making a blunder does a scientist do something which, conceivably, no one else might ever do again. Artists are not troubled by matters of priority, but Wagner would certainly not have spent twenty years on *The Ring* if he had thought it at all possible for someone else to nip in ahead of him with *Götterdämmerung*.

Like other amateurs, Koestler finds it difficult to understand why scientists seem so often to shirk the study of really fundamental or challenging problems. With Robert Graves he regrets the absence of 'intense research' upon variations in the—ah—

'emotive potentials of the sense modalities'. He wonders why 'the genetics of behaviour' should still be 'uncharted territory' and asks whether this may not be because the framework of neo-Darwinism is too rickety to support an enquiry. The real reason is so much simpler: the problem is very, very difficult. Goodness knows how it is to be got at. It may be outflanked or it may yield to attrition, but probably not to direct assault. No scientist is admired for failing in the attempt to solve problems that lie beyond his competence. The most he can hope for is the kindly contempt earned by the Utopian politician. If politics is the art of the possible, research is surely the art of the soluble. Both are immensely practical-minded affairs.

Although much of Koestler's book has to do with explanation, he seems to pay little attention to the narrowly scientific usages of the concept. Some of the 'explanations' he quotes with approval[2] are simply analgesic pills which dull the aches of incomprehension without going to their causes. The kind of explanation the scientist spends most of his time thinking up and testing—the hypothesis which enfolds the matters to be explained among its logical consequences—gets little attention. Instead we are told that there are all kinds of explanations and many degrees of understanding, starting with the '*unconscious* understanding mediated by the dream'.

Dreams bring out the worst in Koestler. Dreaming is a 'sliding back towards the pulsating darkness, one and undivided, of which we were part before our separate egos were formed'. No wonder, then, that the understanding it conveys is of the unconscious kind. 'There is no need to emphasize, in this century of Freud and Jung, that the logic of the dream . . . derives from the magic type of causation found in primitive societies and the fantasies of childhood.' But those who enjoy slopping around in the amniotic fluid should pause for a moment to entertain (perhaps only unconsciously in the first instance) the idea that the content of dreams may be totally devoid of 'meaning'. There should be no need to emphasize, in this century of radio sets and electronic devices, that many dreams may be assemblages of thought-elements that convey no information whatsoever: that they may just be *noise*.[3]

Koestler's theory of the creative act is set out in Book One as a special theory comprehended within a general theory that occupies Book Two. In Book One he defines two special notions, 'matrix' and 'code'. A code is a system of rules of process or performance and a matrix is 'any ability, habit, or skill, any pattern of ordered behaviour' governed by a code. In particular, a matrix of thought is, or can for variety's sake be described as, a 'frame of reference', an 'associative context', a 'type of logic', or a 'universe of discourse'. Behind every act of creation lies a binary association (bisociation) of matrices: in that which provokes laughter they collide, in a new intellectual synthesis they fuse, and in an aesthetic experience they confront each other or are juxtaposed. These three degrees of experience form a continuum, and may indeed grow out of the bisociation of the very same matrices. Thus the exuberant, explosive, tension-relieving delight (*Eureka!*) and the long after-glow ('the oceanic feeling') excited by an intellectual synthesis have their counterparts in laughter and in the sense of satisfaction at seeing a joke.

Koestler is far too intelligent a man not to realize that his account of creative activity is full of difficulties, but though he mentions the contexts in which some of them arise, he does not direct attention to them explicitly or make any attempt to work them out. Among them, and in no special order, are: (*a*) Just how does an explanation which later proves false (as most do: and none, he admits, is proved true) give rise to just the same feelings of joy and exaltation as one which later stands up to challenge? What went wrong: didn't the matrices fuse, or were they the wrong kind of matrix, or what? (*b*) The source of most joy in science lies not so much in devising an explanation as in getting the results of an experiment which upholds it. (*c*) Some awkward problems are raised by the fact that the chap who *sees* a joke splits his sides laughing as well, maybe, as the chap who makes it; but the chap who 'sees' or is apprised of an intellectual synthesis does not share in the tension-relieving, explosive joy of discovery. Likewise the joy of artistic creation 'travels' in some sense in which the joy of intellectual synthesis does not, and this difference between them seems to me to outweigh their similarities. (*d*) It follows from Koestler's scheme that an intellectual synthesis, upon being proved false, should at once become a huge joke, especially to the person who devised it, for it must have rested upon the kind of bisociative act that underlies the comical. However, we are not amused. (*e*) The sense of comfort an explanation may give rise to has nothing to do with bringing about or even witnessing a 'fusion of matrices': laymen get it not so much from knowing an explanation as from knowing that an explanation is known.

I should mention, because Koestler does not, that the so-called 'hypothetico-deductive' interpretation of the scientific process copes perfectly with difficulties (*a*) and (*b*), and that in it (*c*), (*d*) and (*e*) do not arise. Devising a hypothesis is a 'creative act' in the sense that it is the invention of a possible world, or a possible fragment of the world; experiments are then done to find out whether or not that imagined world is, to a good enough approximation, the real one. As Koestler conceives it, the act of creation is not, in the usual sense, creative at all; as he says, it merely 'uncovers, selects, reshuffles, combines, synthesises already existing facts, ideas, faculties, skills'.

Koestler's psychological thought, though not confessedly 'introspective', is in the style of the nineteenth century—a point delicately made, as I read it, by Sir Cyril Burt in his foreword. Koestler nags away at behaviourism, which he describes as 'the dominant school in contemporary psychology', though later he says of J. B. Watson's textbook that 'few students today remember its contents, or even its basic postulates'. For all its crudities behaviourism, conceived as a methodology rather than as a psychological system, taught psychology with brutal emphasis that 'The dog is whining' and 'The dog is sad' are statements of altogether different empirical standing, and heaven help psychology if it ever again overlooks the distinction.

I was dreading the moment in my reading of Koestler when I was to be told that sexual reproduction, 'the bisociation of two genetic codes', was 'the basic model of the creative act'. Koestler's Book Two contains the general theory which comprehends the special theory I have just outlined. It is full of rather old-fashioned biology (but what fun to read again of axial gradients!), and the person who has to be told on page 57 that the sympathetic nervous system has nothing to do with the emotion of sympathy will not make much of it. The argument runs thus. The system of nature is a hierarchical system of elements, sub-elements, sub-sub-elements etc., each enjoying a certain wholeness and autonomy, but each also subordinate to the element above it in the hierarchy. The structure of army command is one example of such a hierarchy (a company has some autonomy but is subordinate to the battalion) and in the living world the hierarchy of organism, organ, cell, cell part etc. is another. At each level we find a certain wholeness and a certain partness.

Koestler declares that at each level of the hierarchy 'homologous' principles operate, with the consequence that any phenomenon at one level must have its homologue or formal counterpart at each other level. In particular, we shall find 'mental equivalents' of what goes on at lower levels in the hierarchy, and conversely, since we can go either up or down the ladder of correspondences, physical equivalents of what goes on in the mind. The 'creative stress' of the artist or scientist corresponds to the 'general alarm reaction' of the injured animal, and dreaming is the mental equivalent of the regenerative processes that make good wear and tear. Embryonic develolpment has a certain self-assertive quality, and so have 'perceptual matrices'. A not yet verbalized analogy corresponds to an organ rudiment in the early embryo, and rhythm and rhyme, assonance and pun, are 'vestigial echoes' of the 'primitive pulsations of living matter'.

No metaphors these: 'they have solid roots in the earth'; but to my ears they sound silly, and I believe them to be as silly as they sound. Disregarding the rights and wrongs of building hierarchies out of non-homogeneous elements (though in fact it won't do to mix up perceptual matrices with adrenal glands, embryos, jokes and rhymes), the correspondences Koestler makes so much of are of the purely formal and abstract kind that can be expressed without any regard to their empirical content. Even if the networks of relationships holding at each level of the hierarchy were isomorphic, there would be no necessary affinity between the things so related. The correspondences which Koestler urges us to believe in are harmless enough, but arguments of this kind can be mischievous (for example, a case for totalitarian government can be built upon an unsound analogy between organism and State).

Koestler makes a good point when he says that during the past hundred years or so scientists have felt themselves under a professional obligation to write in a dry, cold, pulseless way; to be, in short, boring. (It is part of the heritage of inductivism.) *The Act of Creation* is so full of vitality that it creates around it an aura of good-to-be-alive, and though Koestler regards himself as the author of a new and important general psychological theory, I am delighted that, in writing a 'popular' work, he has in a sense appealed to the general public over the heads of the profession. But certain rules of scientific manners must be observed no matter what form the account of a scientific theory may take. One must mention (if only to dismiss with contempt) other, alternative explanations of the matters one is dealing with; and one must discuss (if only to prove them groundless) some of the objections that are likely to be raised against one's theories by the ignorant or ill-disposed. Koestler seems to have no adequate grasp of the importance of *criticism* in science—above all of self-criticism, for most of a scientist's wounds are self-inflicted. Nor can I remember in his book any passage suggesting observations or experiments which might qualify or refine his ideas. He quotes with approval one 'laconic pronouncement' of Dirac's which must have made sense in context but which otherwise sounds just naughty: 'It is more important to have beauty in one's equations than to have them fit experiment.' The high inspirational origin of a theory is no guarantee of its trustworthiness, and Koestler should avoid giving the impression that he thinks it is. No belief could bring science more quickly to ruin.

Mr Koestler replied thus:

SIR—Allow me to answer some of the points raised in Professor Medawar's review of my recent book.

1 Medawar writes:

There are passages in *The Act of Creation* which convince me that he has no real grasp of how scientists go about their work. Consider the depths of misunderstanding revealed by Koestler's aloof and snobbish remarks about the 'unseemly haste' with which some scientists publish their discoveries. Koestler's historical hobnobbings with men of genius seem to have made him forget the fact that, in science, what X misses today Y will surely hit upon tomorrow . . .

The snobbish remarks to which this passage refers read as follows:

In 1922, Ogburn and Thomas published some 150 examples of discoveries and inventions which were made independently by several persons; and, more recently, Merton came to the seemingly paradoxical conclusion that 'the pattern of independent multiple discoveries in science is . . . the dominant pattern rather than a subsidiary one'. He quotes as an example Lord Kelvin, whose published papers contain 'at least thirty-two discoveries of his own which he subsequently found had also been made by others . . .'. The endless priority disputes which have poisoned the supposedly serene atmosphere of scientific research throughout the ages, and the unseemly haste of many scientists to establish priority by rushing into print—or, at least, depositing manuscripts in sealed envelopes with some learned society—point in the same direction. Some—among them Galileo and Hooke—even went to the length of publishing half-completed discoveries in the form of anagrams, to ensure priority without letting rivals in on the idea.

2 Medawar seems to object to my quite unoriginal contention that unconscious processes in the dream and in the hypnagogic state between dreaming and awakening often play a decisive part in scientific discovery. At least this seems to be the meaning behind the heavy veils of irony in the passage:

Dreams bring out the worst in Koestler . . . those who enjoy slopping around in the amniotic fluid should pause for a moment to entertain (perhaps only unconsciously in the first instance) the idea that the content of dreams may be totally devoid of 'meaning' . . . that many dreams may be assemblages of thought-elements that convey no information whatsoever.

No doubt most dreams are self-addressed messages whose information-content is purely private and 'meaningless' to others. But equally undeniable is the fact—which Medawar chooses to pass in silence—that dreams, hypnagogic images and other forms of unconscious intuitions proved decisive in the discoveries of dozens of scientists and mathematicians whose testimonies I quoted—among them Ampère, Gauss, Kekulé, Leibnitz, Poincaré, Fechner, Otto Loewi, Planck, Einstein, to mention only a few.

3 Medawar raises five objections, numbered (*a*) to (*e*), against the theory I proposed. To save space, let me refer to the passages in the book in which the answers to these objections can be found. (*a*) Bk One, Chap. IX, p. 212 et seq. (*b*) The 'joy' in 'devising an explanation' and the satisfaction derived from its empirical confirmation enter at different stages and must not be confused. (*c*) Bk One, Chap. IV, p. 87 et seq. and Chap. XI, p. 255 et seq. (*d*) 'It follows from Koestler's scheme, etc.'. The answer is, it does not. (*e*) See Bk One, Chap. XVII, pp. 325–31.

4 Medawar accuses me of quoting out of context 'one "laconic pronouncement" of Dirac's'. The single sentence which Medawar requotes is on p. 329. The full context, which Medawar overlooked, is to be found on pp. 245–6. If he had no time to read through the book, he should at least have looked at the index.

5 Medawar accuses me of contradicting myself: 'Koestler nags away at behaviourism, which he describes as "the dominant school in contemporary psychology", though later he says of J. B. Watson's textbook that "few students today remember its contents or even its basic postulates".' This I said on p. 558. On p. 559 I continued:

Although the cruder absurdities of Watsonian behaviourism are forgotten, it had laid the foundations on which the later, more refined behaviouristic systems [of Guthrie, Hull,and Skinner] were built; the dominant trend in American and Russian psychology in the generation that followed had a distinctly Pavlov–Watsonian flavour. The methods became more sophisticated, but the philosophy behind them remained the same.

The rest consists of ironic innuendo and *ex cathedra* pronouncements. 'Certain rules of scientific manners must be observed,' Professor Medawar informs us. I wish he had lived up to his precept.

I in turn answered:

I should like to take Mr Koestler's points one by one.

1 *Priority.* A scientist's sense of concern about matters of priority may not be creditable, but only prigs deny its existence, and the fact that it does exist points towards something distinctive in the act of creation as it occurs in science. It is not good enough to brush it aside with clichés ('unseemly haste', 'rushing into print') or to pour scorn on its extremer manifestations. I think my own interpretation was the right one—priority in science gives moral possession—but Koestler seems not to realize that there is anything to interpret. As to simultaneous discovery: it was the slight air of wonderment about it in the very passage Koestler now quotes which made me ask if he realized the consequences of the relationship, peculiar to science, between X and Y. Simultaneous discovery is utterly commonplace, and it was only the rarity of scientists, not the inherent improbability of the phenomenon, that made it remarkable in the past. Scientists on the same road may be expected to arrive at the same destination, often not far apart. Romantics like Koestler don't like to admit this, because it seems to them to derogate from the authority of genius. Thus of Newton and Leibniz, equal first with the differential calculus, Koestler says 'the greatness of this accomplishment is hardly diminished by the fact that two among millions, instead of one among millions, had the exceptional genius to do it'. But millions weren't *trying* for the calculus. If they had been hundreds would have got it. Very simple-minded people think that if Newton had died prematurely we should still be at our wits' end to account for the fall of apples. Is there not just a trace of this *naïveté* in Koestler?

2 *Dreams.* Koestler quite misses the point, and what he says is a good example of how stubbornly the mind may deny entry to the unfamiliar (*The Act of Creation*, p. 216). I did not suggest that dreams conveyed private messages whose import was known only to the dreamer. My proposal, as unoriginal as Koestler's, was that dreams are not messages at all. It is naughty of Koestler to lump together 'dreams, hypnagogic images and other forms of

unconscious intuitions', as if my misgivings about dreams extended to all other unscripted activities of the mind. They don't: if we are to brandish texts at each other, I will cite an article in the *Times Literary Supplement*[4] in which I speak up for inspiration and against the idea that discovery can be logically mechanised.

3 *Objections* (*a*), (*c*) and (*e*). Koestler would not have drawn special attention to these passages in his book unless he really believed them to hold the answers to my, as I think, damaging objections to his theory. Now, on rereading him, I feel convinced that he simply doesn't understand the *kind* of intellectual performance that is expected of someone who propounds or defends a scientific or philosophic theory. But now we have both had our say and it can all go out to arbitration. (*b*) We agree, then: but why here no citations of the passages in his book in which he says so? (*d*) I'm so sorry, but I still think it follows from Koestler's scheme that an intellectual synthesis, upon being proved false, should appear funny. (If I were parodying Koestler I should describe it as a joke played by Nature of which we were very slow to see the point.) What's more, the converse also follows, that a great intellectual synthesis wrongly believed false will be thought hugely comical from the outset. Koestler seems to think so too:

Until the seventeenth century the Copernican hypothesis of the earth's motion was considered as obviously incompatible with common-sense experience; it was accordingly treated as a huge joke . . . The history of science abounds with examples of discoveries greeted with howls of laughter because they seemed to be a marriage of incompatibles . . .[5]

But as I say, in real life the refutation of a hypothesis can be deeply upsetting. If there are howls they are not of laughter.

4 Dirac's allegedly 'laconic pronouncement' ('It is more important to have beauty in one's equations than to have them fit experiment') is in reality a very diffident expression of opinion whose context I was wrong to overlook. But it was rash of Koestler to draw my attention to it, for what Dirac goes on to say is: 'It seems that if one is working from the point of view of getting beauty in one's equation, *and if one has really a sound insight*, one is

on a sure line of progress.' The italics are mine, but Koestler is welcome to them.

5 *Behaviourism*. No, I didn't say Koestler contradicted himself, though I do find his love–hate relationship with modern experimental psychology extremely tiresome. Nor do I think he has quite got my point, which was that even if behaviourism were dead as a system it is still very much alive as a methodology.

Koestler must have had some doubts about the wisdom and taste of his final sentences, and I suppose it will only make matters worse if I say I forgive him. I will not, however, forgive him for hinting that I didn't read his book, nor for the fact that I had to spend hours and hours and hours in doing so.

Postscript

For completeness' sake it should be added that Mr Iain Hamilton in *Koestler: A Biography*[6] published a loyal but in my judgement philosophically inexpert attempt to defend Koestler against my criticisms of *The Act of Creation*. Reconsideration of the whole controversy makes me now more than ever regret that Koestler did not concentrate his attention and his energy upon imaginative writing of the kind in which he is so superbly proficient.

5 Darwin's illness

Charles Darwin was a sick man for the last forty of his seventy-three years of life. His diaries tell the story of a man deep in the shadow of chronic illness—gnawed at by gastric and intestinal pains, frightened by palpitations, weak and lethargic, often sick and shivery, a bad sleeper, and always an attentive student of his own woes. His complaints began about a year after his return from the great scientific adventures that occupied the five-year voyage of HMS *Beagle*, and they soon took on a fitfully recurrent pattern. Over the next few years, as he became progressively weaker, Darwin gave up his more energetic pursuits, including the geological field-work he had until then delighted in; and in 1842, when only thirty-three, he and his devoted wife Emma retired to a country house in Kent. Darwin left Down House seldom and England never, relying upon correspondence to keep himself up with scientific affairs, and in later years looking fearfully upon the hubbub that broke out after the publication of the *Origin of Species* in 1859.

Like many chronic invalids Darwin came to adopt a settled routine—now a little walk, now a little rest, now a little reading—and three or four hours' work a day was about all he could find energy for. Yet he looked well enough, and was very far from being disagreeable. Every account makes him out considerate and loving, and his granddaughter, Gwen Raverat, described him as affectionate, spontaneous and gay. Nor had he a feeble constitution. He had been an open-air man, strongly built, and at Cambridge a keen shot and sportsman. His records of the *Beagle* and subsequently its Master's show him resilient, tough and full of energy. True, he had not been wholly free of illness. At Valparaiso he had had a fever which, Sir Arthur Keith believed, might well have been typhoid (an important point this, for Keith upheld a psychogenic interpretation of Darwin's later illness); and at Plymouth, waiting fretfully for the *Beagle* to set sail, he complained of palpitations and had dark thoughts of heart disease. But nothing about him gave the slightest premonition of the forty years of invalidism that lay ahead.

What was wrong with Darwin? His own doctors were baffled, and their modern descendants disagree. If Darwin's illness had organic signs they were of a kind his doctors could not then have recognized: they inclined to think him a hypochondriac, and the suspicion that they did so is known to have caused Darwin real distress. Orthodox opinion still has it that Darwin's illness was psychogenic, that is, arose from causes in his own mind; indeed, it figures in Alvarez's textbook on the neuroses as a type specimen of neurasthenia. But what lay behind his neurotic illness? Alvarez, after fifty years' reflection on the matter,[1] seemed to think poor heredity answer enough, and drew attention to the number of difficult and eccentric Darwin and Wedgwood relatives (Emma was a Wedgwood and so was Charles's mother). For Professor Hubble,[2] Darwin's illness, though beyond question of emotional origin, was a subtle adaptation which protected him from the rigours and buffetings of everyday life, the demands of society and the public obligations of a great figure in the world of learning. 'Darwin by his psychoneurosis secretly and passionately nourished his genius' and so gave himself time to execute his great scientific labours: he 'could have done his work in no other way'. There is no refuting Hubble's argument, for there is no argument; the case is presented merely by asseveration ('there can be no doubt', 'it is apparent', 'it is inconceivable', 'it is clear', 'there is overwhelming testimony'). Professor Darlington thinks it possible that the persisting cause of Darwin's illness was the disapproval that grew out of Emma's slow recognition that his doctrines were not such as a Christian might approve of.[3] This, however, is a merely casual suggestion; much more weighty is the full-dress psychoanalytic interpretation of Dr Edward Kempf, one to which *The Times* felt its readers' attention should be specially drawn.[4]

Kempf believed that Darwin's forty years' disabling illness was a neurotic manifestation of a conflict between his sense of duty towards a rather domineering father and a sexual attachment to his mother, who died when he was eight. His mother, a gentle and latterly an ailing creature, fond of flowers and pets, had propounded a riddle which it was Darwin's life-work to resolve: How, by looking inside a flower, might its name be discovered? Kempf wrote in 1918 with an arch delicacy that sometimes obscures his meaning, but Good's[5] more recent interpretation leaves us in no doubt. For Good, 'there is a wealth of evidence that unmistakably points' to the idea that Darwin's illness was 'a distorted expression of the aggression, hate, and resentment felt, at an unconscious level, by Darwin towards his tyrannical father'. These deep and terrible feelings found outward expression in Darwin's touching reverence toward his father and his father's memory, and in his describing his father as the kindest and wisest man he ever knew: clear evidence, if evidence were needed, of how deeply his true inner sentiments had been repressed. 'As in the case of Oedipus, Darwin's punishment for the unconscious parricide was a heavy one—almost forty years of severe and crippling neurotic suffering, which left him at his very best fit for a maximum of three hours' daily work.'

It must be made clear that Darwin's father's tyranny was as unconscious as the hatred it gave rise to. Robert Darwin was a very large (340 lb.) and extremely successful Shrewsbury physician who, starting with the £20 given to him by grandfather Erasmus, made a fortune great enough to support all his children in comfort all their lives. He was a rather overbearing man of decided opinions, and we can see an outcrop of his tyrannical inner nature in his reproaching Charles for his idleness and love of sport at Cambridge and for his getting into the company of what Charles called 'dissipated low-minded young men'. Robert also thoroughly disapproved of Charles's ambition to join the *Beagle*, because he had very much wanted him to go, if not into medicine, then into the Church; but later, at least at a conscious level, he withdrew his objections, and it was he who bought Down House for Charles and Emma.

But much else in Darwin's career must have helped to lay the foundations of a lifetime's neurotic illness. Kempf must, I think, have been the first to call attention—obvious though it now seems—to Darwin's intent and continuous preoccupation with matters to do with sex. We need look no further than the titles of his books: the *Origin* itself, of course; *Selection in Relation to Sex*; *The Effects of Cross- and Self-Fertilization in the Vegetable Kingdom*; and *On the Various Contrivances by which Orchids are Fertilized by Insects*. With so great a load of guilt, need we wonder that at the age of thirty-three Darwin should have retired from public life to live in quiet seclusion in the country? It was a sacrificial gesture, even a crucifixion: and Kempf calls attention to the inner significance of the fact that it was at the age of thirty-three that Christ himself was crucified.

What is still more important, I feel, is that psychoanalysis has been able to play a searchlight upon the problem of why Darwin's genius took its distinctive form. Dr Phyllis Greenacre, in her Freud Anniversary Lecture,[6] says she suspects that his turning to science was mainly the consequence of a 'reaction to sadomasochistic fantasies concerning his own birth and his mother's death'. But we can be more particular than this. Kempf reveals to us that when Darwin was speculating upon the selection of favourable variations he was thinking, of course, of Mother's Favourites ('Darwin was unable to avoid unconsciously founding his sincerest conclusions on his own most delicate emotional strivings'); and Good explains how in dethroning his heavenly father Darwin found solace for being unable to slay his earthly one.

What *was* wrong with Darwin? We may never know for certain, and there is no other testimony to overwhelm us, but Professor Saul Adler FRS, of the Hebrew University of Jerusalem, makes a good case[7] for Darwin's having suffered from a chronic and disabling infectious illness called *Chagas' disease* after Carlos Chagas Sr, the distinguished Brazilian medical scientist who first defined it, and caused by a micro-organism whose name, *Trypanosoma cruzi*, honours another distinguished Brazilian scientist, Osvaldo

Cruz. Sir Gavin de Beer, in his excellent *Charles Darwin*,[8] thinks Adler's interpretation by far the most likely one. It differs from other theories we have considered in being based upon the use of reasoning, and Adler's case for it runs approximately thus:

On 26 March 1835, when spending the night in a village in the Argentinian province of Mendoza, Darwin was attacked by the huge blood-sucking bug *Triatoma infestans*, the benchuca. The benchuca, the 'great black bug of the Pampas', is the chief vector of *T. cruzi*; even today more than 60 per cent of the inhabitants of Mendoza give evidence of the disease and 'as many as 70 per cent of specimens of *Triatoma infestans* are infected with the trypanosome'. It is very likely, then, that Darwin was infected: South American experts consulted by Adler put his chances of escape no higher than 'negligible'. The symptomatology of Darwin's illness can, it appears, be matched closely by known cases of Chagas' disease in its chronic form. De Beer summarizes the evidence thus:

The trypanosome invades the muscle of the heart in over 80 per cent of Chagas's disease patients, which makes them very tired; it invades the ganglion cells of Auerbach's nerve plexus in the wall of the intestine, damage to which upsets normal movement and causes great distress; and it invades the auricular-ventricular bundle of the heart which controls the timing of the beats of auricle and ventricle, interference with which may result in heart-block. The lassitude, gastro-intestinal discomfort, and heart trouble from which he suffered an attack in 1873 and died in 1882, all receive a simple and objective explanation if he was massively infected with the trypanosome when he was bitten by the bug on 26 March 1835.

De Beer points out that *T. cruzi* was not identified until twenty-seven years after Darwin's death.

A number of minor and in themselves insubstantial pieces of evidence tell in favour of Adler's interpretation, one of them being that even today inexperienced clinicians may dismiss the chronic form of Chagas' disease as an illness of neurotic origin. I am not aware of any decisive evidence against Adler, and some of the arguments used to discredit his theory establish nothing more than their authors' anxiety to rehabilitate a purely psychogenic interpretation. Disputants so naïve should abstain from public controversy. But clinicians have made it clear to me that the infective theory is by no means a walk-over. There is a general and consistent colouring of hypochondria about Darwin's illness; it is a little surprising that we hear nothing about the acute fever and glandular swellings that would surely have followed infection:[9] and so very long-drawn-out a warfare between host and parasite, neither gaining the upper hand for long, is at least unusual. Adler's interpretation has been widely accepted by Brazilian experts, but I suspect that Darwin's having suffered from an illness so closely associated with the names of two great Brazilian scientists is the source of a certain national pride.

The diagnoses of organic illness and of neurosis are not, of course, incompatible. Human beings cannot be straightforwardly ill like cats and mice; almost all chronic illness is surrounded by a penumbra of gloomy imaginings and by worries and fears that may have physical manifestations. I believe that Darwin was organically ill (the case for his having had Chagas' disease is clearly a strong one) but was also the victim of neurosis; and that the neurotic element in his illness may have been caused by the very obscurity of its origins; by his being 'genuinely' ill, that is to say, and having nothing to show for it—surely a great embarrassment to a man whose whole intellectual life was a marshalling and assay of hard evidence. It is a familiar enough story. Ill people suspected of hypochondria or malingering have to pretend to be iller than they really are and may then get taken in by their own deception. They do this to convince others, but Darwin had also to convince himself, for he had no privileged insight into what was wrong with him. The entries in Darwin's notebooks that bear on his health read to me like the writings of a man desperately reassuring himself of the reality of his illness. 'There,' one can imagine his saying, 'I *am* ill, I must be ill; for how otherwise could I feel like this?'

If this interpretation represents any large part of the truth the physicians who inclined to think Darwin a hypochondriac cannot be held blameless, in spite of the fact that the diagnosis of his ailment, if it was indeed Chagas' disease, was entirely beyond their competence. Even among the tough-minded, the mistaken diagnosis of neurotic illness may cause an extreme exasperation—with symptoms which, of course, serve only to confirm the physician in his diagnosis. But Darwin was a gentle creature who had greatly revered his physician father: to such a man the implied diagnosis of hypochondria would carry special authority and do grave and lasting harm. Perhaps Darwin's physicians should have been more on their guard against an interpretation of his illness that gave him so much less comfort than it gave themselves.

6 Two conceptions of science

My theme is popular misconceptions of scientific thought. I shall argue that the ideas of the educated lay public on the nature of scientific enquiry and the intellectual character of those who carry it out are in a state of dignified, yet utter, confusion. Most of these misconceptions are harmless enough, but some are mischievous, and all help to estrange the sciences from the humanities and the so-called 'pure' sciences from the applied.

Let me begin with an example of what I have in mind. The passage that follows has been made up, but its plaintive sound is so familiar that the reader may find it hard to believe it is not a genuine quotation.

Science is essentially a growth of organized factual knowledge [true or false?], and as science advances, the burden of factual information which it adds to daily is becoming well nigh insupportable. A time will surely come when the scientist must train not for the traditional three or four years, but for ten or more, if he is to equip himself to be a front-line combatant in the battle for knowledge. As things are, the scientist avoids being crushed beneath this factual burden by taking refuge in specialization, and the increase of specialization is the distinguishing mark of modern scientific growth. Because of it, scientists are becoming progressively less well able to communicate even with each other, let alone with the outside world; and we must look forward to an ever finer fragmentation of knowledge, in which each specialist will live in a tiny world of his own. St Thomas Aquinas was the last . . .

True or false, all this? False, I should say, in every particular. Science is no more a classified inventory of factual information than history a chronology of dates. The equation of science with *facts* and of the humane arts with *ideas* is one of the shabby genteelisms that bolster up the humanist's self-esteem. That great Platonist, Goldsworthy Lowes Dickinson, who did his best to keep science below stairs, described Aristotle as 'a man of science in the modern sense' because he was 'a careful collector and observer of an enormous range of facts'. No wonder Lowes Dickinson classified Ideas with Philosophy, Art and Love, but the sciences with—*trade*.

The ballast of factual information, so far from being just about to sink us, is growing daily less. The factual burden of a science varies inversely with its degree of maturity. As a science advances, particular facts are comprehended within, and therefore in a sense annihilated by, general statements of steadily increasing explanatory power and compass—whereupon the facts need no longer be known explicitly, that is, spelled out and kept in mind. In all sciences we are being progressively relieved of the burden of singular instances, the tyranny of the particular. We need no longer record the fall of every apple.

Biology before Darwin was almost all facts. My friend R. B. Freeman has brought to light some Victorian examination questions from our oldest English school of zoology, at University College, London. The answers called for nothing more than a voluble pouring forth of factual information.[1] Certainly there is an epoch in the growth of a science during which facts accumulate faster than theories can accommodate them, but biology is over the hump (though biological learned journals still outnumber learned journals of all other kinds by about three to one); and physics is far enough advanced for an eminent physicist to have assured me, with the air of one not wishing to be overheard, that the science itself was drawing to a close . . .

The case for prolonging a scientist's formal education for many years beyond a humanist's follows naturally from the belief that scientific education is a taking on board of specialized technical knowledge. In real life, the time at which a scientist graduates is less important for scientific than for economic and psychological reasons, and for reasons to do with getting enough people through the universities in good time. The length of university schooling is far more important to those whose education ends with graduation than to those for whom education is an indefinitely continued process.

As to scientists' becoming ever narrower and more specialized: the opposite is the case. One of the distinguishing marks of modern science is the disappearance of sectarian loyalties. Newly graduated biologists have wider sympathies today than they had in my day, just as ours were wider than our predecessors'. At the turn of the century an embryologist could still peer down a microscope into a little world of his own. Today he cannot hope to make head or tail of development unless he draws evidence from bacteriology, protozoology, and microbiology generally; he must know the gist of modern theories of protein synthesis and be pretty well up in genetics.

So it is for biologists generally. Isolationism is over; we all depend upon and sustain each other. I must not speak for specialization in the physical sciences, but feel sure that the continuous and highly successful recruitment of physicists and chemists into biology would not have been possible if they were as specialized as we are often encouraged to believe.

The thoughts I have been criticizing are thus not really thoughts at all, but thought-substitutes, declarations of the kind public people make on public occasions when they are desperately hard up for things to say.

Let me turn now to two serious but completely different conceptions of science, embodying two different valuations of scientific life and of the purpose of scientific enquiry. For dialectical reasons I have exaggerated the differences between them, and I do not suggest that anybody cleaves wholly to the one conception or wholly to the other.

According to the first conception, science is above all else an imaginative and exploratory activity, and the scientist is a man taking part in a great intellectual adventure. Intuition is the mainspring of every advancement of learning, and *having ideas* is the scientist's highest accomplishment; the working out of ideas is an important and exacting but yet a lesser occupation. Pure science requires no justification outside itself, and its usefulness has no bearing on its valuation. 'The first man of science', said Coleridge, 'was he who looked into a thing, not to learn whether it could furnish him with food, or shelter, or weapons, or tools, or ornaments, or *play-withs*, but who sought to know it for the gratification of knowing.'

Science and poetry in its widest sense are cognate, as Shelley so rightly said. So conceived, science can flourish only in an atmosphere of complete freedom, protected from the nagging importunities of need and use, because the scientist must travel where his imagination leads him. Even if a man should spend five years getting nowhere, that might represent an honourable and perhaps even a noble endeavour. The patrons of science—today the Research Councils and the great Foundations—should support men, not projects, and individual men rather than teams, for the history of science is for the most part a history of men of genius.

The alternative conception runs something like this: science is above all else a critical and analytical activity; the scientist is pre-eminently a man who requires evidence before he delivers an opinion, and when it comes to evidence he is hard to please. Imagination is a catalyst merely: it can speed thought but cannot start it or give it direction; and imagination must at all times be under the censorship of a dispassionate and sceptical habit of thought. Science and poetry are antithetical, as Shelley so rightly said.[2] Scientific research is intended to enlarge human understanding, and its usefulness is the only objective measure of the degree to which it does so; as to freedom in science, two world wars have shown us how very well science can flourish under the pressures of necessity. Patrons of science who really know their business will support projects, not people, and most of these projects will be carried out by teams rather than by individuals, because modern science calls for a consortium of the talents and the day of the individual is almost done. If any scientist should spend five years getting nowhere, his ambitions should be turned in some other direction without delay.

I have made the one conception a little more romantic than it really is, and the other a little more worldly, and to restore the balance, I want to express the distinction in a different and, I think, more fundamental way.

In the romantic conception, truth takes shape in the mind of the observer: it is his imaginative grasp of *what might be true* that provides the incentive for finding out, so far as he can, what *is* true. Every advance in science is therefore the outcome of a speculative adventure, an excursion into the unknown. According to the opposite view, truth resides in nature and is to be got at only through the evidence of the senses: apprehension leads by a direct pathway to comprehension, and the scientist's task is essentially one of *discernment*. This act of discernment can be carried out according to a Method which, though imagination can help it, does not depend on the imagination: the Scientific Method will see him through.[3]

Inasmuch as these two sets of opinions contradict each other flatly in every particular, it seems hardly possible that they should both be true; but anyone who has actually done or reflected deeply upon scientific research knows that there is in fact a great deal of truth in both of them. For a scientist must indeed be freely imaginative and yet sceptical, creative and yet a critic. There is a sense in which he must be free, but another in which his thought must be very precisely regimented; there is poetry in science, but also a lot of bookkeeping.

There is no paradox here: it just so happens that what are usually thought of as two alternative and indeed competing accounts of *one* process of thought are in fact accounts of the *two* successive and complementary episodes of thought that occur in every advance of scientific understanding. Unfortunately, we in England have been brought up to believe that scientific discovery turns upon the use of a method analogous to and of the same logical stature as deduction, namely the method of *Induction*—a logically mechanized process of thought which, starting from simple declarations of fact arising out of the evidence of the senses, can lead us with certainty to the truth of general laws. This would be an intellectually disabling belief if anyone actually believed it, and it is one for which John Stuart Mill's methodology of science must take most of the blame. The chief weakness of Millian induction was its failure to distinguish between the acts of mind involved in discovery and in proof. It was an understandable mistake, because in the process of deduction, the paradigm of all exact and conclusive reasoning, discovery and proof may depend on the same act of mind: starting from true premises, we can derive and so 'discover' a theorem by reasoning which (if it has been carried out according to the rules) itself shows that the theorem must be true. Mill thought that his process of 'induction' could fulfil the same two functions; but, alas, mistakenly, for it is not the origin but only the *acceptance* of hypotheses that depends upon the authority of logic.

If we abandon the idea of induction and draw a clear distinction between *having an idea* and *testing it* or *trying it out*—it is as simple as that, though it can be put more grandly—then the antitheses I have been discussing fade away. Obviously 'having an idea' or framing a hypothesis is an imaginative exploit of some kind, the work of a single mind; obviously 'trying it out' must be a ruthlessly critical process to which many skills and many hands may contribute. The form taken by scientific criticism is obvious too: experimentation *is* criticism; that is, experimentation in the modern sense, according to which an experiment is an act performed to test a hypothesis, not in the old Baconian sense, in which an experiment was a contrived experience intended to enlarge our knowledge of what actually went on in nature. Bacon exhorted us, rightly too, not to speculate upon but actually to experiment with loadstone and burning glass and rubbed amber; *his* experiments answer the question 'I wonder what would happen if . . . ?' Baconian experimentation is not a critical activity but a kind of creative play.

The distinction between—and the formal separateness of—the creative and the critical components of scientific thinking is shown up by logical dissection, but it is far from obvious in practice because the two work in a rapid reciprocation of guesswork and checkwork, proposal and disposal, *Conjecture and Refutation*. Though imaginative thought and criticism are equally necessary to a scientist, they are often very unequally developed in any one man. Professional judgement frowns upon extremes. The scientist who devotes his time to showing up the inadequacies of the work of others is suspected of lacking ideas of his own, and everyone soon loses patience with the man who bubbles over with ideas which he loses interest in and fails to follow up.

The general conception of science which reconciles, indeed literally joins together, the two sets of contradictory opinions I have just outlined is sometimes called the 'hypothetico-deductive' conception. For our present clear understanding of the logical structure and wider scientific implications of the hypothetico-deductive system we are of course indebted to Karl Popper's *Logik der Forschung* of 1934, translated into English as *The Logic of Scientific Discovery*.[4]

Everything I have said so far about the hypothetico-deductive system applies with exactly the same force to 'applied' science, even in its simplest and most familiar forms, as to that which is commonly called 'pure' or 'basic'. Imaginative conjecture and criticism, in that order, underlie the physician's diagnosis of his patient's ailments or the mechanic's explanation of why a car won't run. The physician may like to think himself, as Darwin did, an inductivist and a good Baconian, but with equally little reason, for Darwin was no inductivist; no more is he.

What now follows is an attempt to analyse, not the difference between basic and applied science, but the motives which have led people to think it highly important, and above all to make it the basis of an intellectual class-distinction.

Francis Bacon was not the first to distinguish basic from applied science, but no one before him put the matter so clearly and insistently, and the distinction as he draws it is unquestionably just. 'It is an error of special note,' said Bacon, pondering upon the many infirmities of current learning, 'that the industry bestowed upon experiments hath presently, upon the first access into the business, seized upon some designed operation; I mean sought after *Experiments of Use* and not *Experiments of Light and Discovery*.'[5] (The image of light is a favourite of Bacon's, and the idea of kindling a light in nature.) Bacon's distinction is between research that increases our power over nature and research that increases our understanding of nature, and he is telling us that the power comes from the understanding. He felt his distinction upheld by the example of that Divine Builder who created Light only,

and no *Materiate Work*, in the first day, turning to what we
should nowadays presumably call Applied Science in the days
following.

No one now questions Bacon's argument. Who nowadays
would try to build an aeroplane without trying to master the
appropriate aerodynamic theory? Sciences not yet underpinned
by theory are not much more than kitchen arts. Aeronautics, and
the engineering and applied sciences generally, do of course obey
the Baconian ruling that what is done for use should so far as
possible be done in the light of understanding. Unhappily, Bacon's
distinction is not the one we now make when we differentiate
between the basic and applied sciences. The notion of *purity* has
somehow been superimposed upon it, and in a new usage that
connotes a conscious and inexplicably self-righteous disengage-
ment from the pressures of necessity and use. The distinction is
not now between the empirically founded sciences and those
whose axioms were supposedly known a priori; rather it is between polite and rude learning, between the laudably useless and
the vulgarly applied, the free and the intellectually compromised,
the poetic and the mundane.

Let me first say that all this is terribly, terribly English, or
anyhow Anglo-Saxon. Making pure and applied science the basis
of a class distinction helps us to forget that it was our engineers
and merchants, not the armed forces, the Civil Service and the
gentry, who won for us that very grand position in the world from
which we have now stepped down. It is not always easy to explain
to foreigners the whole connotation of 'pure' in the context 'pure
research'. They only shake their heads uneasily and wonder if it
may not have something to do with cricket. They lack also our
Own Very Special Blend of high-mindedness and humbug in that
reasoning which champions Pure Research because, while it en-
ables the human spirit to breathe freely in the thin and serene
atmosphere of the intellectual highlands, it is also a splendid long-
term investment. Invest in applied science for quick returns (the
spiritual message runs), but in pure science for capital appreciation.[6] And so we make a special virtue of encouraging pure
research in, say, cancer institutes or institutes devoted to the study

of rheumatism or the allergies—always in the hope, of course, that the various lines of research, like the lines of perspective, will converge somewhere upon a point. But there is nothing virtuous about it! We encourage pure research in these situations because we know no other way to go about it. If we knew of a direct pathway leading to the solution of the clinical problem of rheumatoid arthritis, can anyone seriously believe that we should not take it?

The more creditable part of our English reverence for pure research derives, I believe, from a certain accident of our aesthetic history. Let us concede that imaginative thought plays an important part—no matter what—in discovery and invention. Now in this country the quintessential form of imaginative activity has always been poetic invention. Hereabouts, a man inspired is typically a poet inspired. Unfortunately there is no such thing as Applied Poetry—or rather, there is, but we think little of it. We look askance at poetry for the occasion, even for Royal occasions. For poetry 'is not like reasoning, a power to be exerted according to the determination of the will. A man cannot say "I will compose poetry." The greatest poet even cannot say it.' Still less can he say that he will compose joyful or lugubrious poetry, or poetry upon a given theme. 'Poetry . . . is not subject to the control of the active powers of the mind, and its birth and recurrence have no necessary connection with the consciousness or will.'

Substitute 'pure science' for 'poetry' in Shelley's manifesto, and it will help us to understand the aesthetic conspiracy which has led us to think so much more highly of pure research than of research with an acknowledged practical purpose. It was quite otherwise in the early days of the Royal Society, when there was a danger that any experiment not immediately useful would be dismissed as play.

It is strange [said Thomas Sprat][7] that we are not able to inculcate into the minds of many men, the necessity of that distinction of my Lord Bacon's, that there ought to be Experiments of Light, as well as of Fruit . . . If they will persist in contemning all Experiments, except those that bring with them immediate gain, and a present Harvest, they may as well cavil at the Providence of God, that he has not made all the seasons of the year to be times of mowing, reaping, and vintage.

Sprat is not arguing for pure research in the sense in which we should now use that term, but rather against a hasty opportunism; his formula is 'Light now, for Use hereafter'.

In countries in which poetry is not the top art form, the idea of occasional or commissioned art is commonplace and honourable, and there is correspondingly less fuss, if any, about the distinction between pure science and applied. Tapestry and statuary, stained glass, murals and portraiture, palaces, cathedrals and town halls, fire music, water music and funeral marches—most are commissioned, and the act of being commissioned may itself light up the imagination.[8] For everyone who uses imagination knows that it can be trained and guided and deliberately stocked with things to be imaginative about. Only the irremediably romantic can believe, as Coleridge did, that artistic creation is a microcosmic version of that Divine sort of creation which can make something out of nothing, or out of a homogeneous cloud of forms or notions—and how little right *he* had to think so has been made clear by *The Road to Xanadu* and other aetiological studies of Coleridge's choice of images and words. To the sober-minded the 'spontaneity' of an idea signifies nothing more than our unawareness of what preceded its irruption into conscious thought.

I am labouring these obvious points in order to make it clear that poetic inspiration is not a valid guide to imaginative activity in all its forms; that there is no case for looking down on commissioned art or science or on extra-mural sources of inspiration; and that our English reverence for Pure Research, though historically understandable, and perhaps even lovable, is also slightly ridiculous.

If we study the criteria that underlie a scientist's own valuation of science, we shall certainly not find purity among them. This is a significant omission, for the scientific valuation of scientific research is remarkably uniform throughout the world.

Here then are some of the criteria used by scientists when judging their colleagues' discoveries and the interpretations put upon them. Foremost is their *explanatory value*—their rank in the grand hierarchy of explanations and their power to establish new pedigrees of research and reasoning. A second is their clarifying power, the degree to which they resolve what has hitherto been perplexing; a third, the feat of originality involved in the research, the surprisingness of the solution to which it led, and so on. Scientists give weight (though much less weight than mathematicians do) to the elegance of a solution and the economy of the thought and work that went into it; they give credit, too, for the difficulty of the enterprise as a whole—the size of the obstacles that had to be got over or got round before the solution was reached. But purity, as such, is nowhere. Nor is usefulness, which has its own scale of valuation and its own rewards. Let usage guide us: 'How neat!' one scientist might say of another's work— or 'How ingenious!'—or 'How very illuminating!'—but never, in my hearing anyway, 'How pure!'

There is without doubt a case for uncommitted or disengaged research, but it is not self-evident, and there may turn out to be better ways of doing what pure research professes to do. For example, in institutes of basic research it is believed and hoped that something practically useful may be come upon in the course of free-ranging enquiry, whereupon research which has hitherto shed diffuse light will now come sharply into focus. This procedure works; that is, it works sometimes, and it may be the best we can do, but there's no knowing, for alternative approaches have not yet been tried out on a sufficiently large scale. Might not the converse approach be equally effective, given equal opportunity and equal talent?—to start with a concrete problem, but then to allow the research to open out in the direction of greater generality, so that the more particular and special discoveries can be made to rank as theorems derived from statements of higher explanatory value. I can see no reason why this approach, *if it were to be attempted by persons of the same ability*, should not work just as well as its more conventional alternative; in fact I believe that some great American companies are moving towards it and already have some brilliant achievements to justify their choice— for example, the growth of a generalized communications theory out of the practical problems of sending messages by telephone.

Research done in this style is always in focus, and those who carry it out, if temporarily baffled, can always retreat from the general into the particular.

If our reverence for Pure Science is a rather parochial thing, a by-product of the literary propaganda of the romantic revival; if no case can be made for it on philosophic grounds; if purity is not part of a scientist's own valuation of science; then why on earth do we think so highly of it? It is, I think, our humanist brethren who have taught us to believe that, while pure science is a genteel and even creditable activity for scientists in universities, applied science, with all its horrid connotations of trade, has no place on the campus; for only the purest of pure science can give countenance to research in the humanities—research which, though it cannot very well be described as pure, for want of anything applied to compare it with, can all too readily be described as useless. The humanist fears that if we abandon the ideal of pure knowledge, knowledge acquired for its own sake, then usefulness becomes the only measure of merit; and that if it does become so, research in the humane arts is doomed.

These fears, I have tried to explain, are groundless. Neither its purity *nor* its usefulness enter a scientist's valuation of his own research. The scientist values research by the size of its contribution to that huge, logically articulated structure of ideas which is already, though not yet half built, the most glorious accomplishment of mankind. The humanist must value his research by different but equally honourable standards, particularly by the contribution it makes, directly or indirectly, to our understanding of human nature and conduct, and human sensibility.

I have been trying to make a case for a critical study of the organization of research, by which I do *not* mean either the allocation of administrative responsibilities for research or the economics and logistics of science,[9] important though they are. I mean a study of the behavioural and intellectual structure of everything that goes into the enlargement of our knowledge and understanding of nature. I have already mentioned a few of the problems that might be on the agenda of such an investigation, and here are a few more. Are scientists a homogeneous body of people in respect of temperament, motivation, and style of thought? (Obviously not: but we talk of *the* scientist nevertheless.) Is there such a thing as a 'scientific mind'? I think not. Or *the* scientific method? Again, I think not. What exactly are the terms of a scientist's contract with the truth? This is an important question, for according to the interpretation of the scientific process which I myself think the most plausible, a scientist, so far from being a man who never knowingly departs from the truth, is always *telling stories* in a sense not so very far removed from that of the nursery euphemism—stories which might be about real life but which must be tested very scrupulously to find out if indeed they are so.[10]

Again, and in no particular order: Is it really true that a good or genuine scientist is, or should be, indifferent to matters of priority, caring only for the Advancement of Learning and nothing for who causes it to come about? How can the *frettoso* of research be combined harmoniously with the *adagio* of administration? How can the productivity of scientists be increased: is full-time research really a good thing for more than a lucky or slightly obsessional minority, and, if not, what else should a scientist do, and how should his time be parcelled out to best advantage?

These are important questions, and their answers must no longer be entrusted to asseveration—to 'peremptory fits of asseveration', Bacon said, when clearing the ground for his own *Great Instauration*. They will have to be thought over and argued out with some sense of urgency; and we here in England had better be quick about it, in case the wind changes and we get fixed permanently in our Anglo-Saxon attitudes to research.

7 Science and the sanctity of life

I do not intend to deny that the advance of science may sometimes have consequences that endanger, if not life itself, then the
quality of life or our self-respect as human beings (for it is in this
wider sense that I think 'sanctity' should be construed). Nor shall
I waste time by defending science as a whole or scientists generally against a charge of inner or essential malevolence. The
Wicked Scientist is not to be taken seriously: Dr Strangelove, Dr
Moreau, Dr Moriarty, Dr Mabuse, Dr Frankenstein (an honorary
degree, this) and the rest of them are puppets of Gothic fiction.
Scientists, on the whole, are amiable and well-meaning creatures.
There must be very few wicked scientists. There are, how-
ever, plenty of wicked philosophers, wicked priests and wicked
politicians.

One of the gravest charges ever made against science is that
biology has now put it into our power to corrupt both the body
and the mind of man. By scientific means (the charge runs) we
can now breed different kinds and different races—different
'makes', almost—of human beings, degrading some, making
aristocrats of others, adapting others still to special purposes:
treating them in fact like dogs, for this is how we *have* treated
dogs. Or again: science now makes it possible to dominate and
control the thought of human beings—to improve them, perhaps,
if that should be our purpose, but more often to enslave or to
corrupt with evil teaching.

But these things have always been possible. At any time in the
past five thousand years it would have been within our power to
embark on a programme of selecting and culling human beings

and raising breeds as different from one another as toy poodles and Pekinese are from St Bernards and Great Danes. In a genetic sense the empirical arts of the breeder are as easily applicable to human beings as to horses—more easily applicable, in fact, for human beings are highly *evolvable* animals, a property they owe partly to an open and uncomplicated breeding system, which allows them a glorious range of inborn diversity and therefore a tremendous evolutionary potential; and partly to their lack of physical specializations (in the sense in which ant-eaters and woodpeckers and indeed dogs are specialized), a property which gives human beings a sort of amateur status among animals. And it has always been possible to pervert or corrupt human beings by coercion, propaganda or evil indoctrination. Science has not yet improved these methods, nor have scientists used them. They have, however, been used to great effect by politicians, philosophers and priests.

The mischief that science may do grows just as often out of trying to do good—as, for example, improving the yield of soil is intended to do good—as out of actions intended to be destructive. The reason is simple enough: however hard we try, we do not and sometimes cannot foresee all the distant consequences of scientific innovation. No one clearly foresaw that the widespread use of antibiotics might bring about an evolution of organisms resistant to their action. No one could have predicted that X-irradiation was a possible cause of cancer. No one could have foreseen the speed and scale with which advances in medicine and public health would create a problem of over-population that threatens to undo much of what medical science has worked for. (Thirty years ago the talk was all of how the people of the Western world were reproducing themselves too slowly to make good the wastage of mortality; we heard tell of a 'Twilight of Parenthood', and wondered rather fearfully where it all would end.) But somehow or other we shall get round all these problems, for every one of them is soluble, even the population problem, and even though its solution is obstructed above all else by the bigotry of some of our fellow men.

I choose from medicine and medical biology one or two concrete examples of how advances in science threaten or seem to threaten the sanctity of human life. Many of these threats, of course, are in no sense distinctively medical, though they are often loosely classified as such. They are merely medical contexts for far more pervasive dangers. One of them is our increasing state of dependence on medical services and the medical industries. What would become of the diabetic if the supplies of insulin dried up, or of the victims of Addison's disease deprived of synthetic steroids? Questions of this kind might be asked of every service that society provides. In a complex society we all sustain and depend upon each other—for transport, communications, food, goods, shelter, protection and a hundred other things. The medical industries will not break down all by themselves, and if they do break down it will be only one episode of a far greater disaster.

The same goes for the economic burden imposed by illness in any community that takes some collective responsibility for the health of its citizens. All shared burdens have a cost which is to a greater or lesser degree shared between us: education, pensions, social welfare, legal aid and every other social service, including government.

We are getting nearer what is distinctively medical when we ask ourselves about the economics, logistics and morality of keeping people alive by medical intervention and medical devices. At present it is the cost and complexity of the operation, and the shortage of machines and organs, that denies a kidney graft or an artificial kidney to anyone mortally in need of it. The limiting factors are thus still economic and logistic. But what about the morality of keeping people alive by these heroic medical contrivances? I do not think it is possible to give any answer that is universally valid or that, if it were valid, would remain so for more than a very few years. Medical contrivances extend all the way from pills and plasters and bottles of tonic to complex mechanical prostheses, which will one day include mechanical hearts. At what point shall we say we are wantonly interfering with nature and prolonging life beyond what is proper and humane?

In practice the answer we give is founded not upon abstract moralizing but upon a certain natural sense of the fitness of things, a feeling that is shared by most kind and reasonable people even if we cannot define it in philosophically defensible or legally accountable terms. It is only at international conferences that we tend to adopt the convention that people behave like idiots unless acting upon clear and well-turned instructions to behave sensibly. There is in fact no general formula or smooth form of words we can appeal to when in perplexity.

Moreover, our sense of what is fit and proper is not something fixed, as if it were inborn and instinctual. It changes as our experience grows, as our understanding deepens, and as we enlarge our grasp of possibilities—just as living religions and laws change, and social structures and family relationships.

I feel that our sense of what is right and just is already beginning to be offended by the idea of taking great exertions to keep alive grossly deformed or monstrous newborn children, particularly if their deformities of body or mind arise from major defects of the genetic apparatus. There are in fact scientific reasons for changing an opinion that might have seemed just and reasonable a hundred years ago.

Everybody takes it for granted, because it is so obviously true, that a married couple will have children of very different kinds and constitutions on different occasions. But the traditional opinion, which most of us are still unconsciously guided by, is that the child conceived on any one occasion is the unique and necessary product of that occasion: *that* child would have been conceived, we tend to think, or no child at all. This interpretation is quite false, but human dignity and security clamour for it. A child sometimes wonderingly acknowledges that he would never have been born at all if his mother and father had not chanced to meet and fall in love and marry. He does not realize that, instead of conceiving him, his parents might have conceived any one of a hundred thousand other children, all unlike each other and unlike himself. Only over the past one hundred years has it come to be realized that the child conceived on any one occasion belongs to a vast cohort of Possible Children, any one of whom might have been conceived and born if a different spermatozoon had chanced to fertilize the mother's egg cell—and the egg cell itself is only one of very many. It is a matter of luck then, a sort of genetic lottery. And sometimes it is cruelly bad luck—some terrible genetic conjunction, perhaps, which once in ten or twenty thousand times will bring together a matching pair of damaging recessive genes. Such a misfortune, being the outcome of a random process, is, considered in isolation, completely and essentially pointless. It is not even strictly true to say that a particular inborn abnormality must have lain within the genetic potentiality of the parents, for the malignant gene may have arisen *de novo* by mutation. The whole process is unhallowed—is, in the older sense of that word, profane.[1]

I am saying that if we feel ourselves under a moral obligation to make every possible exertion to keep a monstrous embryo or new born child alive *because* it is in some sense the naturally in- tended—and therefore the unique and privileged—product of its parents' union at the moment of its conception, then we are making an elementary and cruel blunder: for it is *luck* that deter- mines which one child is in fact conceived out of the cohort of Possible Children that might have been conceived by those two parents on that occasion. I am not using the word 'luck' of conception as such, nor of the processes of embryonic and foetal growth, nor indeed in any sense that derogates from the wonder and awe in which we hold processes of great complexity and natural beauty which we do not fully understand; I am simply using it in its proper sense and proper place.[2]

This train of thought leads me directly to eugenics—'the science', to quote its founder, Francis Galton, 'which deals with all the influences that improve the inborn qualities of a race; also with those that develop them to the utmost advantage'. Because the upper and lower boundaries of an individual's capability and performance are set by his genetic make-up, it is clear that if eugenic policies were to be ill-founded or mistakenly applied they could offer a most terrible threat to the sanctity and dignity of human life. This threat I shall now examine.

Eugenics is traditionally subdivided into positive and negative eugenics. Positive eugenics has to do with attempts to improve human beings by genetic policies, particularly policies founded upon selective or directed breeding. Negative eugenics has the lesser ambition of attempting to eradicate as many as possible of our inborn imperfections. The distinction is useful and pragmatically valid for the following reasons.[3] Defects of the genetic constitution (such as those which manifest themselves as Down's syndrome ('mongolism'), haemophilia, galactosaemia, phenylketonuria and a hundred other hereditary abnormalities) have a much simpler genetic basis than desirable characteristics like beauty, high physical performance, intelligence or fertility. This is almost self-evident. All geneticists believe that 'fitness' in its most general sense depends on a nicely balanced co-ordination and interaction of genetic factors, itself the product of laborious and long drawn out evolutionary adjustment. It is inconceivable, indeed self-contradictory, that an animal should evolve into the possession of some complex pattern of interaction between genes that made it inefficient, undesirable, or unfit—that is, *less* well adapted to the prevailing circumstances. Likewise, a motor car will run badly for any one of a multitude of particular and special reasons, but runs well because of the harmonious mechanical interactions made possible by a sound and economically viable design.

Negative eugenics is a more manageable and understandable enterprise than positive eugenics. Nevertheless, many well-meaning people believe that, with the knowledge and skills already available to us, and within the framework of a society that upholds the rights of individuals, it is possible in principle to raise a superior kind of human being by a controlled or 'recommended' scheme of mating and by regulating the number of children each couple should be allowed or encouraged to have. If stockbreeders can do it, the argument runs, why should not we?—for who can deny that domesticated animals have been improved by deliberate human intervention?

I think this argument is unsound for a lesser and for a more important reason.

1 Domesticated animals have not been 'improved' in the sense in which we should use that word of human beings. They have not enjoyed an all-round improvement, for some special characteristics or faculties have been so far as possible 'fixed' without special regard to and sometimes at the expense of others. Tameness and docility are most easily achieved at the expense of intelligence, but that does not matter if what we are interested in is, say, the quality and yield of wool.

2 The ambition of the stockbreeder in the past, though he did not realize it, was twofold: not merely to achieve a predictably uniform product by artificial selection, but also to establish an internal genetic uniformity (homozygosity) in respect of the characters under selection, to make sure that the stock would 'breed true'—for it would be a disaster if characters selected over many generations were to be irrecoverbly mixed up in a hybrid progeny. The older stockbreeder believed that uniformity and breeding true were characteristics that necessarily went together, whereas we now know that they can be separately achieved. And he expected his product to fulfil two quite distinct functions which we now know to be separable, and often better separated: on the one hand, to be in themselves the favoured stock and the top performers—the super-sheep or super-mice—and, on the other hand, to be the parents of the next generation of that stock. It is rather as if Rolls-Royces, in addition to being an end-product of manufacture, had to be so designed as to give rise to Rolls-Royce progeny.[4]

It is just as well these older views are mistaken, for with naturally outbreeding populations such as our own, genetic uniformity, arrived at and maintained by selective inbreeding, is a highly artificial state of affairs with many inherent and ineradicable disadvantages.

Stockbreeders, under genetic guidance, are now therefore inclining more and more towards a policy of deliberate and nicely calculated cross-breeding. In the simplest case, two partially inbred and internally uniform stocks are raised and perpetuated to provide two uniform lineages of parents, but the eugenic goal, the marketable end-product or high performer, is the progeny of a cross between members of the two parental stocks. Being of hybrid make-up, the progeny do not breed true, and are not in fact bred from; they can be likened to a manufactured end-product: but they can be uniformly reproduced at will by crossing the two parental stocks. Many more sophisticated regimens of cross-breeding have been adopted or attempted, but the innovation of principle is the same. (1) The end-products are all like each other and are faithfully reproducible, but are not bred from because they do not breed true: the organisms that represent the eugenic goal have been relieved of the responsibility of reproducing themselves. And (2) the end-products, though uniform in the sense of being like each other, are to a large extent hybrid—heterozygous as opposed to homozygous—in genetic composition.

The practices of stockbreeders can therefore no longer be used to support the argument that a policy of positive eugenics is applicable in principle to human beings in a society respecting the rights of individuals. The genetical manufacture of supermen by a policy of cross-breeding between two or more parental stocks is unacceptable today, and the idea that it might one day become acceptable is unacceptable also.

A deep fallacy does in fact eat into the theoretical foundations of positive eugenics and that older conception of stockbreeding out of which it grew. The fallacy was to suppose that the *product* of evolution, that is the outcome of an episode of evolutionary change, was a new and improved genetic formula (genotype) which conferred a higher degree of adaptedness on the individuals that possessed it. This improved formula, representing a new and more successful solution of the problems of remaining alive in a hostile environment, was thought to be shared by nearly all members of the newly evolved population, and to be stable except in so far as further evolution might cause it to change again. Moreover, the population would have to be predominantly homozygous in respect of the genetic factors entering into the new formula, for otherwise the individuals possessing it would not breed true to type, and everything natural selection had won would be squandered in succeeding generations.

Most geneticists think this view mistaken. It is *populations* that evolve, not the lineages and pedigrees of old-fashioned evolutionary 'family trees', and the end-product of an evolutionary episode is not a new genetic formula enjoyed by a group of similar individuals, but a new spectrum of genotypes, a new pattern of genetic inequality, definable only in terms of the population as a whole. Naturally outbreeding populations are not genetically uniform, even to a first approximation. They are persistently and obstinately diverse in respect of nearly all constitutive characters which have been studied deeply enough to say for certain whether they are uniform or not. It is the *population* that breeds true, not its individual members. The progeny of a given population are themselves a population with the same pattern of genetic make-up as their parents—except in so far as evolutionary or selective forces may have altered it. Nor should we think of uniformity as a desirable state of affairs which *we* can achieve even if nature, unaided, cannot. It is inherently undesirable, for a great many reasons.

The goal of positive eugenics, in its older form, cannot be achieved, and I feel that eugenic policy must be confined (paraphrasing Karl Popper) to *piecemeal genetic engineering*. That is just what negative eugenics amounts to; and now, rather than to deal in generalities, I should like to consider a concrete eugenic problem and discuss the morality of one of its possible solutions.

Some 'inborn' defects—some defects that are the direct consequence of an individual's genetic make-up as it was fixed at the moment of conception—are said to be of *recessive* determination. By a recessive defect is meant one that is caused by, to put it crudely, a 'bad' gene that must be present in *both* the gametes that unite to form a fertilized egg, that is, in both spermatozoon and egg cell, not just in one or the other. If the bad gene *is* present in only one of the gametes, the individual that grows out of its fusion with the other is said to be a *carrier* (technically, a heterozygote).

Recessive defects are individually rather rare—their frequency is of the order of 10^{-4} (one in ten thousand)—but collectively they are most important. Among them are, for example, phenylketonuria, a congenital inability to handle a certain dietary constituent, the amino acid phenylalanine, a constituent of many proteins; galactosemia, another inborn biochemical deficiency, the victims of which cannot cope metabolically with galactose, an immediate derivative of milk sugar; and, more common than either, fibrocystic disease of the pancreas, believed to be the symptom of a generalized disorder of mucus-secreting cells. All three are caused by particular genetic defects; but their secondary consequences are manifold and deep-seated. The phenylketonuric baby is on the way to becoming an imbecile. The victim of galactosemia may become blind through cataract and be mentally retarded.

Contrary to popular superstition, many congenital ailments can be prevented or, if not prevented, cured. But in this context prevention and cure have very special meanings.

The phenylketonuric or galactosemic child may be protected from the consequences of his genetic lesion by keeping him on a diet free from phenylalanine in the one case or lactose in the other. This is a most unnatural proceeding, and much easier said than done, but I take it no one would be prepared to argue that it was an unwarrantable interference with the workings of providence. It is not a cure in the usual medical sense because it neither removes nor repairs the underlying congenital deficiency. What it does is to create around the patient a special little world, a microcosm free from phenylalanine or galactose as the case may be, in which the genetic deficiency cannot express itself outwardly.

Now consider the underlying morality of prevention. We can prevent phenylketonuria by preventing the genetic conjunction responsible for it in the first instance, that is, by preventing the coming together of an egg cell and a sperm each carrying that same one harmful recessive gene. All but a very small proportion of overt phenylketonurics are the children of parents who are both carriers—carriers, you remember, being the people who inherited the gene from one only of the two gametes that fused at their conception. Carriers greatly outnumber the overtly afflicted. When two carriers of the same gene marry and bear children, one-quarter of their children (on the average) will be normal, one-quarter will be afflicted, and one-half will be carriers like themselves. We shall accomplish our purpose, therefore, if, having identified the carriers (another thing easier said than done, but it *can* be done, and in an increasing number of recessive disorders), we try to discourage them *from marrying each other* by pointing out the likely consequences if they do so. The arithmetic of this is not very alarming. In a typical recessive disease, about one marriage in every five or ten thousand would be discouraged or warned against, and each disappointed party would have between fifty and a hundred other mates to choose from.

If this policy were to be carried out, the overt incidence of a disease like phenylketonuria, in which carriers can be identified, would fall almost to zero between one generation and the next.

Nevertheless the first reaction to such a proposal may be one of outrage. Here is medical officiousness planning yet another insult to human dignity, yet another deprivation of the rights of man. First it was vaccination and then fluoride; if now people are not to be allowed to marry whom they please, why not make a clean job of it and overthrow the Crown or the United States Constitution?

But reflect for a moment. What is being suggested is that a certain small proportion of marriages should be discouraged for genetic reasons, to help us do our best to avoid bringing into the world children who are biochemically crippled. In all cultures marriages are already prohibited for genetic reasons—the prohibition, for example, of certain degrees of inbreeding (the exact degree varies from one culture or religion to another). Thus the prohibition of marriage has an immemorial authority behind it. As to the violation of human dignity entailed by performing tests on engaged couples that are no more complex or offensive than blood tests, let me say only this: if anyone thinks or has ever thought that religion, wealth or colour are matters that may properly be taken into account when deciding whether or not a certain marriage is a suitable one, then let him not dare to suggest that the genetic welfare of human beings should not be given equal weight.

I think that engaged couples should themselves decide, and I am pretty certain they would be guided by the thought of the welfare of their future children. When it came to be learned, about twenty years ago, that marriages between Rhesus-positive men and Rhesus-negative women might lead to the birth of children afflicted by haemolytic disease, a number of young couples are said to have ended their engagements—needlessly, in most cases, because the dangers were overestimated through not being understood. But that is evidence enough that young people marrying today are not likely to take a stand upon some hypothetical right to give birth to defective children, if, by taking thought, they can do otherwise.

The problems I have been discussing illustrate very clearly the way in which scientific evidence bears upon decisions that are not, of course, in themselves scientific. If the termination of a pregnancy is now in question, scientific evidence may tell us that the chances of a defective birth are 100 per cent, 50 per cent, 25 per cent, or perhaps unascertainable. The evidence is highly relevant to the decision, but the decision itself is not a scientific one, and I see no reason why scientists as such should be specially well qualified to make it. The contribution of science is to have enlarged beyond all former bounds the evidence we must take account of before forming our opinions. Today's opinions may not be the same as yesterday's, because they are based on fuller or better evidence. We should quite often have occasion to say 'I used to think that once, but now I have come to hold a rather different opinion.' People who never say as much are either ineffectual or dangerous.

We all nowadays give too much thought to the material blessings or evils that science has brought with it, and too little to its power to liberate us from the confinements of ignorance and superstition.

It may be that the greatest liberation of thought ever achieved by the scientific revolution was to have given mankind the expectation of a future in this world. The idea that the world has a virtually indeterminate future is comparatively new. Much of the philosophic speculation of three hundred years ago was oppressed by the thought that the world had run its course and was coming shortly to an end.[5] 'I was borne in the last Age of the World,' said John Donne, giving it as the 'ordinarily received' opinion that the world had thrice two thousand years to run between its creation and the Second Coming. According to Archbishop Ussher's chronology more than five and a half of those six thousand years had gone by already.[6]

No empirical evidence challenged this dark opinion. There were no new worlds to conquer, for the world was known to be spherical and therefore finite; certainly it was not all known, but the full extent of what was *not* known was known. Outer space did not put into people's minds then, as it does into ours now, the idea of a tremendous endeavour just beginning.

Moreover, life itself seemed changeless. The world a man saw about him in adult life was much the same as it had been in his own childhood, and he had no reason to think it would change in his own or his children's lifetime. We need not wonder that the promise of the next world was held out to believers as an inducement to put up with the incompleteness and inner pointlessness of this one: the present world was only a staging post on the way to better things. There was a certain awful topicality about Thomas Burnet's description of the world in flames at the end of its long journey from 'a dark chaos to a bright star', for the end of the world might indeed come at any time. And Thomas Browne warned us against the folly and extravagance of raising monuments and tombs intended to last for many centuries. We are living in The Setting Part of Time, he told us: *the Great Mutations of the World are acted: it is too late to be ambitious.*

Science has now made it the ordinarily received opinion that the world has a future reaching beyond the most distant frontiers of the imagination—and that is perhaps why, in spite of all his faults, so many scientists still count Francis Bacon their first and greatest spokesman: we may yet build a New Atlantis. The point is that when Thomas Burnet exhorted us to become 'Adventurers for Another World' *he* meant the next world—but we mean this one.

8 J.B.S.

The lives of scientists, considered as Lives, almost always make dull reading. For one thing, the careers of the famous and the merely ordinary fall into much the same pattern, give or take an honorary degree or two, or (in European countries) an honorific order. It could hardly be otherwise. Academics can only seldom lead lives that are spacious or exciting in a worldly sense. They need laboratories or libraries and the company of other academics. Their work is in no way made deeper or more cogent by privation, distress or worldly buffetings. Their private lives may be unhappy, strangely mixed up or comic, but not in ways that tell us anything special about the nature or direction of their work. Academics lie outside the devastation area of the literary convention according to which the lives of artists and men of letters are intrinsically interesting, a source of cultural insight in themselves. If a scientist were to cut his ear off, no one would take it as evidence of a heightened sensibility; if a historian were to fail (as Ruskin did) to consummate his marriage, we should not suppose that our understanding of historical scholarship had somehow been enriched.

The lives of writers, however, are thought to give out a low rumble of cultural portents. One day in the summer of 1968 *The Times* of London devoted a whole column on its front page to new discoveries about the circumstances under which Joseph Conrad came to be discharged by the captain of the *Riversdale*, in which he was serving as first mate—discoveries described as 'extremely important' for the understanding of Conrad and his art, because traces of the incident are to be discerned in Conrad's fiction. (But what are we to make of a scale of values in which such a discovery ranks as extremely important? Would it not have been of epoch-making significance—nay, downright interesting—if Conrad had *not* used his experience of shipboard life in writing his stories about the sea?)

Yet J. B. S. Haldane's life, as Ronald Clark recounts it,[1] is fascinating from end to end. Unless one is in the know already, there is no foretelling at one moment what comes next. Haldane had a flying start in life. His father was a famous physiologist; his uncle translated Schopenhauer and became Lord Chancellor and Minister of War; his aunt was a distinguished social reformer; and his sister Naomi Mitchison (the dedicatee, incidentally, of *The Double Helix*) is a well-known writer and, among other things, an honorary member of the Bakgatla tribe. Even the house he was brought up in has been transformed into an Oxford College.

The Dragon School, Eton and Oxford gave Haldane about the best education a man of his generation could have. At Oxford he was an authentic 'Double First', for having taken first-class honours in Mathematical Moderations, he switched to philosophy and ancient history and took a first-class degree in Greats, the most prestigious thing an Oxford undergraduate could do. Haldane could have made a success of any one of half a dozen careers—as mathematician, classical scholar, philosopher, scientist, journalist or imaginative writer. In unequal proportions he was in fact all of these. On his life's showing he could not have been a politician, administrator (heavens, no!), jurist or, I think, a critic of any kind. In the outcome he became one of the three or four most influential biologists of his generation.

In some respects—quickness of grasp, and the power to connect things in his mind in completely unexpected ways—he was the cleverest man I ever knew. He had something novel and theoretically illuminating to say on every scientific subject he chose to give his mind to: on the kinetics of enzyme action, on infectious disease as a factor in evolution, on the relationship between antigens and genes, and on the impairment of reasoning by prolonged exposure to high concentrations of carbon dioxide. Haldane was the first to describe the genetic phenomenon of linkage in animals generally, and the first to estimate the mutation rate in man. His greatest work began in the 1920s, when independently of Sewall Wright and R. A. Fisher he undertook to refound Darwinism upon the concepts of Mendelian genetics. It should have caused a great awakening of Darwinian theory, and in due course it did so; but at the time it did no more than make Darwinism stir in its dogmatic slumbers, and even today, on the Continent, what passes for Darwinism is essentially the Darwinism of fifty years ago. (The same goes for the neo-Darwinism denounced by modern nature-philosophers, who are handicapped by the fact that the mathematical theory of natural selection is too difficult for them to understand.)

The gist of the newer evolutionism is as follows. In principle, every living organism can be allotted a formula representing its genetic constitution, its make-up in terms of 'genes'. In a natural population of some one species, each gene represented in the population will have a certain frequency of occurrence, because it will often occur in some members but not in others, and the population considered as a whole will therefore itself have a certain genetic make-up, definable in terms of the frequency of genes. In 1908 the mathematician G. H. Hardy announced the following fundamental theorem: that although, through the workings of Mendelian heredity, a virtually infinite variety of new combinations and reassortments of genes will turn up in members of the population, yet if mating is random, and all organisms have an equal chance of leaving offspring, the frequency of each gene in the population will necessarily remain constant from one generation to the next. We must therefore look to some impressed 'force' acting over the population as a whole if the gene frequency is to change in some systematic (that is, other than merely random) way—if, in short, the population is to evolve. The most important of these forces is natural selection, a compendious name for all the agencies that cause one section of the population to make a disproportionately large contribution to the ancestry of future generations. The rate of change of gene frequency is therefore a measure of the magnitude of the force of natural selection.

In the light of this new conception, Haldane, Fisher and Wright were able for the first time to describe the phenomenon of evolution in a genetic language, and to reveal the delicacy and subtlety of a population's response to selective forces. In this scheme of thinking, 'mutation' occupies a certain special place. Mutation adds to genetic diversity, and therefore enlarges the candidature for evolutionary change. This is quite different from saying (as older Darwinians and modern nature-philosophers say) that the mutant organism is itself the candidate for evolution, the hopeful variant that is selected or rejected as the case may be. The newer evolution theory represents a revolution of thought of the same general kind and the same stature as that which led to the development of statistical mechanics in the latter half of the nineteenth century.

This then is 'classical' work, assimilated into all the standard textbooks. If he had done nothing else, Haldane would still be classified as one of the grand masters of modern evolution theory. Yet he was not a profoundly original thinker. His genius was to enrich the soil, not to bring new land into cultivation. He was not himself the author of any great new biological conception, nor did his ideas arouse the misgivings and resentment so often stirred up by what is revolutionary or profoundly original. On the contrary, everything he said was at once recognized as fruitful and illuminating, something one would have been proud and delighted to have thought of oneself, even if later research should prove it to be mistaken.

Haldane had a reputation for being a bad experimenter, anyhow in the narrower, manipulative sense. I never saw him in the traditional laboratory uniform of white or once-white coat, and he gave the impression of being clumsy with his hands. He could design experiments, of course, and guide others in their execution, but he was not by nature an experimentalist; he did not translate scientific problems into a language of conjecture and refutation. His great strength was to see connections, to put two and two together, to work out the deeper or remoter consequences of taking certain theoretical views. If he had been a physicist he would have been a theoretical physicist (and in a small way he actually was), but experimentalists liked talking to him ('Let's try it out on the Prof.'), and only the obtuse could have failed to derive benefit from what he said.

Between his First in Greats in 1914 and his return to Oxford in 1919 to study physiology, Haldane was away at the wars. I knew Haldane only during the latter half of his life, and had not realized until I read Clark's biography how thoroughly Haldane enjoyed everything that went with war. The First World War seems to have been the happiest period of Haldane's life; we have his word for it that he disliked Eton, and he was not yet victim of the many vexations, real or imagined, that took the edge off his enjoyment of professional life. Haldane described life in the front line as 'truly enviable'; he enjoyed the comradeship of war and even (if what he says is anything to go by) the experience of killing people: 'I get a definitely enhanced sense of life when my life is in moderate danger.' Courage he disclaimed, but he was to all appearances fearless. His bravery, as I construe it, was the product of a superb intellectual arrogance—a complete confidence in the accuracy of his assessment of degrees of risk. Like Houdini, he judged safe the exploits that to others seemed suicidal: 'I once bicycled across a gap in full view of the Germans, having foreseen that they would be too surprised to open fire.' When the Germans started using chlorine in 1915, Haldane was taken out of the front line to help his father study its physiological effects—research of the utmost importance which showed up Haldane at his very best, and carried out at some risk (which he underestimated) to himself and his physiological colleagues.

What are we to make of Haldane as a human being? The first thing to be said in answer to such a question is that we are under no obligation to make anything of him at all. It makes no difference now. It might have made a difference if Haldane in his lifetime could have been made to realize the degree to which his work was obstructed by his own perversity. He was so ignorant of anything to do with administration that he did not even know how to call the authorities' attention to the contempt in which he held them. When he burst into terrible anger about his grievances, it was over the heads of minor functionaries and clerks. The cleaners were terrified of him, and the electricians were said to have demanded danger money for working in his room. His room was therefore never cleaned; it became a sort of showpiece, littered with fossil specimens undergoing a second interment.

Clark describes a scene which those who knew him came to regard as typical. On behalf of one of his students Haldane applied for one of the Agricultural Research Council's postgraduate awards. These awards are made provisionally, and are confirmed if the candidate gets a high enough degree. To speed things, one of the Council's junior officers rang up Haldane's secretary to find out what class of degree the candidate had in fact been given. As it happened, the class lists had not yet been published, so the reasonable answer would have been 'I'm sorry, we can't tell you yet because the results aren't out.' Instead Haldane accused the Council of blackmail and an attempt to violate the secrecy of exams: 'I refuse to give you the information, and withdraw my request for a grant. I shall pay for her out of my own pocket.' Yet more than once he scored an important victory: over the *Sex Viri*, for example, a sort of *buffo* male-voice sextet that tried to deprive him of his readership at Cambridge on the grounds of immorality. Indeed, the scenes accompanying the divorce that freed Charlotte Burghes to become Haldane's first wife read like the libretto of a comic opera, including an adultery that was chaste in spite of all appearances to the contrary.

In America, Haldane was notorious for his communist professions: he was ideologically a communist during the latter part of his life, joining the party officially in 1942 and leaving it furtively around 1948, though he continued to write for the *Daily Worker* until 1950. Clark quotes a draft letter of resignation from the party written in 1948, but perhaps never posted. The reasons he gave for wanting to leave the party were so utterly trivial—a squabble about royalties, and various accusations of bad faith—as to make one question the seriousness and solemnity of the motives that led him to join it in the first place. In his public professions

Haldane was the complete party man. Lysenko he thought 'a very fine biologist', so Clark tells us; and I know from conversation with him that he thought it quite likely that Beria, then lately disgraced, had been in the pay of the Americans, and that Slansky and Clementis, the victims of ritual hangings in Czechoslovakia, had got the punishment they deserved.

People were wont to ask how such a clever man could be so completely taken in by Communist propaganda, but Haldane was not clever in respect of any faculty that enters into political judgement. He was totally lacking in worldly sense, a sulky innocent, a whole-hearted believer in *Them*—the agents of that hidden conspiracy against ordinary decent people, the authorities who withheld the grants he had never asked for and who broke the promises they had never made.

We must not take all Haldane's protestations at their face value. His declaration that he left England to live in India because of the disgrace of Suez was an effective way of expressing his contempt for the Suez adventure, but it simply wasn't true. I remember Haldane's once going back on a firm promise to chair a lecture given by a distinguished American scientist on the grounds that it would be too embarrassing for the lecturer: he had once been the victim of a sexual assault by the lecturer's wife. The accusation was utterly ridiculous, and Haldane did not in the least resent my saying so. He didn't want to be bothered with the chairmanship, and could not bring himself to say so in the usual way. But the trouble was that his extravagances became self-defeating. He became a 'character', and people began laughing in anticipation of what he would say or be up to next. It is a sort of Anglo-Saxon form of liquidation, more humane but politically not much less effective than the form of liquidation he condoned. In the Russia of Haldane's day, as Mr Clark makes clear, Haldane would have been much more offensive and with very much better reasons, but he would not have lasted anything like so long.

Physical bravery, but sometimes moral cowardice; intelligence and folly, reasonableness and obstinacy, kindness and aggressiveness, generosity and pettiness—it is like a formulary for all mankind: Haldane was a Johnsonian figure, a with-knobs-on variant of us all; but unless we are bored with life or altogether fed up with human beings, we shall not tire of reading how lofty thoughts can go with silly opinions, or how a man may strive for freedom and yet sometimes condone the work of its enemies.

9 Lucky Jim

On 30 May 1953 James Watson and Francis Crick published in *Nature* a correct interpretation of the crystalline structure of deoxyribonucleic acid, DNA. It was a great discovery, one which went far beyond merely spelling out the spatial design of a large, complicated and important molecule. It explained how that molecule could serve genetic purposes—that is to say, how DNA, within the framework of a single common structure, could exist in forms various enough to encode the messages of heredity. It explained how DNA could be stable in a crystalline sense and yet allow for mutability. Above all it explained in principle, at a molecular level, how DNA undergoes its primordial act of reproduction, the making of more DNA exactly like itself. The great thing about their discovery was its completeness, its air of finality. If Watson and Crick had been seen groping towards an answer; if they had published a partly right solution and had been obliged to follow it up with corrections and glosses; if the solution had come out piecemeal instead of in a blaze of understanding: then it would still have been a great episode in biological history, but something more in the common run of things; something splendidly well done, but not done in the grand romantic manner.

The work that ended by making biological sense of the nucleic acids began forty years ago in the shabby laboratories of the Ministry of Health in London. In 1928 Dr Fred Griffith, one of the Ministry's Medical Officers, published in the *Journal of Hygiene* a paper describing strange observations on the behaviour of pneumococci—behaviour which suggested that they could undergo something akin to a transmutation of bacterial species. The pneumococci exist in a variety of genetically different 'types', distinguished one from another by the chemical make-up of their outer sheaths. Griffith injected into mice a mixture of dead pneumococcal cells of one type and living cells of another type, and in due course he recovered living cells of the type that distinguished the dead cells in the original mixture. On the face of it, he had observed a genetic transformation. There was no good reason to question the results of the experiment. Griffith was a well known and highly expert bacteriologist whose whole professional life had been devoted to describing and defining the variant forms of bacteria, and his experiments (which forestalled the more obvious objections to the meaning he read into them) were straightforward and convincing. Griffith, above all an epidemiologist, did not follow up his work on pneumococcal transformation; nor did he witness its apotheosis, for in 1941 a bomb fell in Endell Street which blew up the Ministry's laboratory while he and his close colleague William Scott were working in it.

The analysis of pneumococcal transformations was carried forward by Martin Dawson and Richard Sia in Columbia University and by Lionel Alloway at the Rockefeller Institute. Between them they showed that the transformation could occur during cultivation outside the body, and that the agent responsible for the transformation could pass through a filter fine enough to hold back the bacteria themselves. These experiments were of great interest to bacteriologists because they gave a new insight into matters having to do with the ups and downs of virulence; but most biologists and geneticists were completely unaware that they were in progress. The dark ages of DNA came to an end in 1944 with the publication from the Rockefeller Institute of a paper by Oswald Avery and his young colleagues, Colin MacLeod and Maclyn McCarty, which gave very good reasons for supposing that the transforming agent was 'a highly polymerized and viscous form of sodium deoxyribonucleate'. This interpretation aroused much resentment, for many scientists unconsciously deplore the resolution of mysteries they have grown up with and have therefore come to love. It nevertheless withstood all efforts to unseat it. Geneticists marvelled at its significance, for the agent that brought about the transformation could be thought of as a naked gene. So very probably the genes were not proteins after all, and the nucleic acids themselves could no longer be thought of as a sort of skeletal material for the chromosomes.

The new conception was full of difficulties, the most serious being that (compared with the baroque profusion of different kinds of proteins) the nucleic acids seemed too simple in make-up and too little variegated to fulfil a genetic function. These doubts were set at rest by Crick and Watson: the combinatorial variety of the four different bases that enter into the make-up of DNA is more than enough to specify or code for the twenty different kinds of amino acids of which proteins are compounded; more than enough, indeed, to convey the detailed genetic message by which one generation of organisms specifies the inborn constitution of the next. Thanks to the work of Crick and half a dozen others, the form of the genetic code, the scheme of signalling, has now been clarified, and thanks to work to which Watson has made important contributions, the mechanism by which the genetic message is mapped into the structure of a protein is now in outline understood.

It is simply not worth arguing with anyone so obtuse as not to realize that this complex of discoveries is the greatest achievement of science in the twentieth century. I say 'complex of discoveries' because discoveries are not a single species of intellection; they are of many different kinds, and Griffith's and Crick-and-Watson's were as different as they could be. Griffith's was a synthetic discovery, in the philosophic sense of that word. It did not close up a visible gap in natural knowledge, but entered upon territory not until then known to exist. If scientific research had stopped by magic in, say, 1920, our picture of the world would not be *known* to be incomplete for want of it. The elucidation of the structure of DNA was analytical in character. Ever since W. T. Astbury published his first X-ray diffraction photographs we all knew that DNA had a crystalline structure, but until the days of Crick and Watson no one knew what it was. The gap was visible then, and if research had stopped in 1950 it would be visible still; our picture of the world would be known to be imperfect. The importance of Griffith's discovery was historical only (I do not mean this in a depreciatory sense). He might not have made it; it might not have been made to this very day; but if he had not, then some other, different discovery would have served an equivalent purpose, that is, would in due course have given away the genetic function of DNA. The discovery of the structure of DNA was logically necessary for the further advance of molecular genetics. If Watson and Crick had not made it, someone else would certainly have done so—almost certainly Linus Pauling, and almost certainly very soon. it would have been that same discovery, too; nothing else could take its place.

Watson and Crick (so Watson tells us) were extremely anxious that Pauling should *not* be the first to get there. In one uneasy hour they feared he had done so, but to their very great relief his solution was erroneous, and they celebrated his failure with a toast. Such an admission will shock most laymen: so much, they will feel, for the 'objectivity' of science; so much for all that fine talk about the disinterested search for truth. In my opinion the idea that scientists ought to be indifferent to matters of priority is simply humbug. Scientists are entitled to be proud of their accomplishments, and what accomplishments can they call 'theirs' except the things they have done or thought of first? People who criticize scientists for wanting to enjoy the satisfaction of intellectual ownership are confusing possessiveness with pride of possession. Meanness, secretiveness and sharp practice are as much despised by scientists as by other decent people in the world of ordinary everyday affairs; nor, in my experience, is generosity less common among them, or less highly esteemed.

It could be said of Watson that, for a man so cheerfully conscious of matters of priority, he is not very generous to his predecessors. The mention of Astbury is perfunctory and of Avery a little condescending. Fred Griffith is not mentioned at all. Yet a paragraph or two would have done it without derogating at all from the splendour of his own achievement. Why did he not make the effort?

It was not lack of generosity, I suggest, but stark insensibility.

These matters belong to scientific history, and the history of science bores most scientists stiff. A great many highly creative scientists (I classify Jim Watson among them) take it quite for granted, though they are usually too polite or too ashamed to say so, that an interest in the history of science is a sign of failing or unawakened powers. It is not good enough to dismiss this as cultural barbarism, a coarse renunciation of one of the glories of humane learning. It points towards something distinctive about scientific learning, and instead of making faces about it we should try to find out why such an attitude is natural and understandable. A scientist's present thoughts and actions are of necessity shaped by what others have done and thought before him; they are the wave-front of a continuous secular process in which The Past does not have a dignified independent existence on its own. Scientific understanding is the integral of a curve of learning; science therefore in some sense comprehends its history within itself. No Fred, no Jim: that is obvious, at least to scientists; and being obvious it is understandable that it should be left unsaid. (I am speaking, of course, about the history of scientific endeavours and accomplishments, not about the history of scientific ideas. Nor do I suggest that the history of science may not be profoundly interesting as history. What I am saying is that it does not often interest the scientist as science.)

Jim Watson ('James' doesn't suit him) majored in zoology in Chicago and took his Ph.D. in Indiana, aged twenty-two. When he arrived in Cambridge in 1951 there could have been nothing much to distinguish him from any other American 'postdoctoral' in search of experience abroad. By 1953 he was world-famous. How much did he owe to luck?

The part played by luck in scientific discovery is greatly overrated. *Ces hasards ne sont que pour ceux qui jouent bien*, as the saying goes. The paradigm of all lucky accidents in science is the discovery of penicillin—the spore floating in through the window, the exposed culture plate, the halo of bacterial inhibition around the spot on which it fell. What people forget is that Fleming had been *looking* for penicillin, or something like it, since the middle of the First World War. Phenomena such as these will not be appreciated, may not be knowingly observed, except against a background of prior expectations. A good scientist is discovery-prone. (As it happens there *was* an element of blind luck in the discovery of penicillin, though it was unknown to Fleming. Most antibiotics—hundreds are now known—are murderously toxic, because they arrest the growth of bacteria by interfering with metabolic processes of a kind that bacteria have in common with higher organisms. Penicillin is comparatively innocuous because it happens to interfere with a synthetic process peculiar to bacteria, namely the synthesis of a distinctive structural element of the bacterial cell wall.)

I do not think Watson was lucky except in the trite sense in which we are all lucky or unlucky—that there were several branching points in his career at which he might easily have gone off in a direction other than the one he took. At such moments the reasons that steer us one way or another are often trivial or ill thought-out. In England a schoolboy of Watson's precocity and style of genius would probably have been steered towards literary studies. It just so happens that during the 1950s, the first great age of molecular biology, the English Schools of Oxford and particularly of Cambridge produced more than a score of graduates of quite outstanding ability—much more brilliant, inventive, articulate and dialectically skilful than most young scientists; right up in the Watson class. But Watson had one towering advantage over all of them: in addition to being extremely clever he had something important to be clever *about*. This is an advantage which scientists enjoy over most other people engaged in intellectual pursuits, and they enjoy it at all levels of capability. To be a first-rate scientist it is not necessary (and certainly not sufficient) to be extremely clever, anyhow in a pyrotechnic sense. One of the great social revolutions brought about by scientific research has been the democratization of learning. Anyone who combines strong common sense with an ordinary degree of imaginativeness can become a creative scientist, and a happy one besides, in so far as happiness depends upon being able to develop to the limit of one's abilities.

Lucky or not, Watson was a highly privileged young man.

Throughout his formative years he worked first under and then with scientists of great distinction; there were no dark unfathomed laboratories in his career. Almost at once (and before he had done anything to deserve it) he entered the privileged inner circle of scientists among whom information is passed by a sort of beating of tom-toms, while others await the publication of a formal paper in a learned journal. But because it was unpremeditated we can count it to luck that Watson fell in with Francis Crick, who (whatever Watson may have intended) comes out in this book as the dominant figure, a man of very great intellectual powers. By all accounts, including Watson's, each provided the right kind of intellectual environment for the other. In no other form of serious creative activity is there anything equivalent to a collaboration between scientists, which is a subtle and complex business, and a triumph when it comes off, because the skill and performance of a team of equals can be more than the sum of individual capabilities. It was a relationship that did work, and in doing so brought them the utmost credit.

Considered as literature, *The Double Helix* will be classified under Memoirs, Scientific. No other book known to me can be so described. It will be an enormous success, and deserves to be so—a classic in the sense that it will go on being read. As with all good memoirs, a fair amount of it consists of trivialities and idle chatter. Like all good memoirs it has not been emasculated by considerations of good taste. Many of the things Watson says about the people in his story will offend them, but his own artless candour excuses him, for he betrays in himself faults graver than those he professes to discern in others. *The Double Helix* is consistent in literary structure. Watson's gaze is always directed outward. There is no philosophizing or psychologizing to obscure our understanding; Watson displays but does not observe himself. Autobiographies, unlike all other works of literature, are part of their own subject-matter. Their lies, if any, are lies *of* their authors but not *about* their authors, who (when discovered in falsehood) merely reveal a truth about themselves, namely that they are liars. Although it sounds a bit too well remembered, Watson's scientific narrative strikes me as perfectly convincing. This is not

to say that the apportionments of credits or demerits are necessarily accurate: that is something which cannot be decided in
abstraction, but only after the people mentioned in the book have
had their say, if they choose to have it. Nor will an intelligent
reader suppose that Watson's judgements upon the character,
motives and probity of other people (sometimes apparently
shrewd, sometimes obviously petty) are 'true' simply because he
himself believes them to be so.

A good many people will read *The Double Helix* for the insight
they hope it will bring them into the nature of the creative process
in science. It may indeed become a standard case history of the so-
called 'hypothetico-deductive' method at work. Hypothesis and
inference, feedback and modified hypothesis, the rapid alter-
nation of imaginative and critical episodes of thought—here it can
all be seen in motion, and every scientist will recognize the
same intellectual structure in the research he does himself. It is
characteristic of science at every level, and indeed of most exploratory or investigative processes in everyday life. No layman
who reads this book with any kind of understanding will ever
again think of the scientist as a man who cranks a machine of
discovery. No beginner in science will henceforward believe that
discovery is bound to come his way if only he practises a certain
Method, goes through a certain well-defined performance of hand
and mind.

Nor, I hope, will anyone go on believing that The Scientist is
some definite kind of person. Given the context, one could not
plausibly imagine a collection of people more different in origin
and education, in manner, manners, appearance, style and
worldly purposes than the men and women who are the characters in this book. Watson himself and Crick and Wilkins, the
central figures; Dorothy Crowfoot and poor Rosalind Franklin,
the only one of them not then living; Perutz, Kendrew and
Huxley; Todd and Bragg, at that time holder of 'the most prestigious chair in science'; Pauling *père et fils*; Bawden and Pirie, in a
momentary appearance; Chargaff; Luria; Mitchison and Griffith
(John, not Fred)—they come out larger than life, perhaps, and
as different one from another as Caterpillar and Mad Hatter.

Watson's childlike vision makes them seem like the creatures of a Wonderland, all at a strange contentious noisy tea-party which made room for him because for people like him, at this particular kind of party, there is always room.

Postscript

'Lucky Jim' was a defence of Watson against the storm of out-raged criticism that burst out after the publication of *The Double Helix*. Nothing has occurred to shake my belief that the discovery of the structure and biological functions of the nucleic acids is the greatest achievement of science in the twentieth century. In de-fending Watson I felt much as advocates must feel when defend-ing a client who is unmistakably guilty of many of the charges brought against him: I have in mind particularly his lack of ad-equate acknowledgement of the work of scientists such as Chargaff who made really important contributions to the eluci-dation of the problem which he and Crick finally solved. I showed Francis Crick my review before it appeared in the *New York Review of Books* and was very pleased when he said of my parallel with Alice's Wonderland and the Mad Hatter's tea-party: 'That's quite right, you know, it was exactly like that.'

One passage in this review of *The Double Helix* came in for a lot of criticism. I see it struck W. H. Auden[1] too, unfavourably I should guess. The passage (p. 99 above) runs:

It just so happens that during the 1950s, the first great age of molecular biology, the English schools of Oxford and particularly of Cambridge produced more than a score of graduates of quite outstanding ability— much more brilliant, inventive, articulate and dialectically skilful than most young scientists; right up in the Jim Watson class. But Watson had one towering advantage over all of them: in addition to being extremely clever he had something important to be clever *about*.

Surely, I was asked, you don't intend to imply that Shakespeare and Tolstoy etc. are not important and that it is hardly possible to be clever about them? Of *course* this is not what I intended. I had it in mind that many of the brilliant contemporaries of Jim Watson and many of the brightest literary students of the later

1950s entered the advertising or entertainment industries or contented themselves with petty literary pursuits. The widely prevalent opinion that almost any literary work, even if it amounts to no more than writing advertising copy or a book review, not to mention that Ph.D. thesis on 'Some little known laundry bills of George Moore', is intrinsically superior to almost any scientific activity is not one with which a scientist can be expected to sympathize.

10 On 'the effecting of all things possible'

1

My title, or, if you like, my motto, comes from Francis Bacon's *New Atlantis*, published in 1627. The *New Atlantis* was Bacon's dream of what the world might have been, and might still become, if human knowledge were directed towards improving the worldly condition of man. It makes a rather strange impression nowadays, and very few people bother with it who are not interested either in Bacon himself, or in the flux of seventeenth-century opinion or the ideology of Utopias. We shall not read it for its sociological insights, which are non-existent, nor as science fiction, because it has a general air of implausibility; but there is one high poetic fancy in the *New Atlantis* that stays in the mind after all its fancies and inventions have been forgotten. In the New Atlantis, an island kingdom lying in very distant seas, the only commodity of external trade is—*light*: Bacon's own special light, the light of understanding. The Merchants of Light who carry out its business are members of a society or order of philosophers who between them make up (so their spokesman declares) 'the noblest foundation that ever was upon the earth'. 'The end of our foundation', the spokesman goes on to say, 'is the knowledge of causes and the secret motions of things; and the enlarging of the bounds of human empire, to the effecting of all things possible.' You will see later on why I chose this motto.

2

My purpose is to draw certain parallels between the spiritual or philosophic condition of thoughtful people in the seventeenth century and in the contemporary world, and to ask why the great philosophic revival that brought comfort and a new kind of understanding to our predecessors has now apparently lost its power to reassure us and cheer us up.

The period of English history that lies roughly between the accession of James I in 1603 and the English Civil War has much in common with the present day.[1] For the historian of ideas, it is a period of questioning and irresolution and despondency; of sermonizing but also of satire; of rival religions competing for allegiance, among them the 'black doctrine of absolute reprobation'; a period during which our human propensity towards hopefulness was clouded over by a sense of inconstancy and decay. Literary historians have spoken of a 'metaphysical shudder',[2] and others of a sense of crisis or of a 'failure of nerve'.[3] Of course, we must not imagine that ordinary people went around with the long sunk-in faces to be expected in the victims of a spiritual deficiency disease. It was philosophic or reflective man who had these misgivings, the man who is all of us some of the time but none of us all of the time, and we may take it that, then as now, the remedy for discomforting thoughts was less often to seek comfort than to abstain from thinking.

Amidst the philosophic gloom of the period I am concerned with, new voices began to be heard which spoke of hope and of the possibility of a future (a subject I shall refer to later on); which spoke of confidence in human reason, and of what human beings might achieve through an understanding of nature and a mastery of the physical world. I think there can be no question that, in this country, it was Francis Bacon who started the dawn chorus—the man who first defined the newer purposes of learning and, less successfully, the means by which they might be fulfilled. Human spirits began to rise. To use a good old seventeenth-century metaphor, there was a slow change, but ultimately a complete one, in the 'climate of opinion'. It became no longer the thing to mope. In a curious way the Pillars of Hercules—the 'Fatal Columns' guarding the Straits of Gibraltar that make the frontispiece to Bacon's *Great Instauration*—provided the rallying cry of the New Philosophy. Let me quote a great

American scholar's, Dr Marjorie Hope Nicolson's,[4] description of how this came about:

> Before Columbus set sail across the Atlantic, the coat of arms of the Royal Family of Spain had been an *impressa*, depicting the Pillars of Hercules, the Straits of Gibraltar, with the motto, *Ne Plus Ultra*. There was 'no more beyond'. It was the glory of Spain that it was the outpost of the world. When Columbus made his discovery, Spanish Royalty thriftily did the only thing necessary: erased the negative, leaving the Pillars of Hercules now bearing the motto, *Plus Ultra*. There was more beyond . . .

And so *plus ultra* became the motto of the New Baconians, and the frontispiece to the *Great Instauration* shows the Pillars of Hercules with ships passing freely to and fro.

One symptom of the new spirit of enquiry was, of course, the foundation of the Royal Society and of sister academies in Italy and France. That story has often been told, and in more than one version, because the parentage of the Royal Society is still in question.[5] We shall be taking altogether too narrow a view of things, however, if we suppose that the great philosophic uncertainties of the seventeenth century were cleared up by the fulfilment of Bacon's ambitions for science. Modern scientific research began earlier than the seventeenth century.[6] The great achievement of the latter half of the seventeenth century was to arrive at a general scheme of belief within which the cultivation of science was seen to be very proper, very useful, and by no means irreligious. This larger conception or purpose, of which science was a principal agency, may be called 'rational humanism' if we are temperamentally in its favour and take our lead from the writings of John Locke, or 'materialistic rationalism' if we are against it and frown disapprovingly over Thomas Hobbes, but neither description is satisfactory, because the new movement had not yet taken on the explicit character of an alternative or even an antidote to religion, which is the sense that 'rational humanism' tends to carry with it today.

However we may describe it, rational humanism became the dominant philosophic influence in human affairs for the next 150 years, and by the end of the eighteenth century the spokesmen of Reason and Enlightenment—men such as Adam Ferguson and

William Godwin and Condorcet—take completely for granted many of the ideas that had seemed exhilarating and revolutionary in the century before. But over this period an important transformation was taking place. The seventeenth-century doctrine of the *necessity* of reason was slowly giving way to a belief in the *sufficiency* of reason—so illustrating the tendency of many powerful human beliefs to develop into an extreme or radical form before they lose their power to persuade us, and in doing so to create anew many of the evils for which at one time they professed to be the remedy. (It has often been said that rationalism in its more extreme manifestations could only supplant religion by acquiring some of the characteristics of religious belief itself.) Please don't interpret these remarks as any kind of attempt to depreciate the power of reason. I emphasize the distinction between the ideas of the necessity and of the sufficiency of reason as a defence against that mad and self-destructive form of anti-rationalism which seems to declare that because reason is not sufficient, it is not necessary.

Many reflective people nowadays believe that we are back in the kind of intellectual and spirtual turmoil that disturbed the first half of the seventeenth century. Both epochs are marked, not by any characteristic system of beliefs (neither can be called 'The Age of' anything), but by an equally characteristic syndrome of unfixed beliefs; by the emptiness that is left when older doctrines have been found wanting and none has yet been found to take their place. Both epochs have the characteristics of a philosophic interregnum. In the first half of the seventeenth century, the essentially medieval world-picture of Elizabethan England had lost its power to satisfy and bring comfort, just as nowadays the radical materialism traditionally associated with Victorian thinkers seems quite inadequate to remedy our complaints. By a curious inversion of thinking, scholastic reasoning is said to have failed because it discouraged new enquiry, but that was precisely the measure of its success. For that is just what successful, satisfying explanations do: they confer a sense of finality; they remove the incentive to work things out anew. At all events the repudiation of Aristotle and the hegemony of ancient learning, of the scholastic style of reasoning, of the illusion of a Golden Age, is as commonplace in the writings of the seventeenth century as dismissive references to rationalism and materialism in the literature of the past fifty years.

We can draw quite a number of detailed correspondences between the contemporary world and the first forty or fifty years of the seventeenth century, all of them part of a syndrome of dissatisfaction and unbelief; and though we might find reason to cavil at each one of them individually, they add up to an impressive case. Novels and philosophical *belles-lettres* have now an inward-looking character, a deep concern with matters of personal salvation and a struggle to establish the authenticity of personal existence; and we may point to the prevalence of satire and of the Jacobean style of 'realism' on the stage. I shall leave aside the political and economic correspondences between the two epochs,[7] important though they are, and confine myself to analogies that might be described as 'philosophical' in the homely older sense, the sense that has to do with the purpose and conduct of life and with the attempt to answer the simple questions that children ask. Once again we are oppressed by a sense of decay and deterioration, but this time, in part at least, by a fear of the deterioration of the world through technological innovation. Artificial fertilizers and pesticides are undermining our health (we tell ourselves), soil and sea are being poisoned by chemical and radioactive wastes, drugs substitute one kind of disease for another, and modern man is under the influence of stimulants whenever he is not under the influence of sedatives. Once again there is a feeling of despondency and incompleteness, a sense of doubt about the adequacy of man, amounting in all to what a future historian might again describe as a failure of nerve. Intelligent and learned men may again seek comfort in an elevated kind of barminess (but something kind and gentle nevertheless). Mystical syntheses between science and religion, like the Cambridge Neoplatonism of the mid-seventeenth century, have their counterpart today, perhaps, in the writings and cult of Teilhard de Chardin and in a revival of faith in the Wisdom of the East. Once again there is a rootlessness or ambivalence about philosophical thinking, as if the discovery or rediscovery of the insufficiency of reason had given a paradoxical validity to nonsense, and this gives us a special sympathy for the dilemmas of the seventeenth century. To William Lecky, the great nineteenth-century historian of rationalism, it seemed almost beyond comprehension that witch hunting and witch burning should have persisted far into the seventeenth century, or that Joseph Glanvill should have been equally an advocate of the Royal Society and of belief in witchcraft.[8]

We do not wonder at it now. It no longer seems strange to us that Pascal the geometer who spoke with perfect composure about infinity and the infinitesimal should have been supplanted by Pascal the great cosmophobe who spoke with anguish about the darkness and loneliness of outer space. Discoveries in astronomy and cosmology have always a specially disturbing quality. We remember the dismay of John Donne and Pascal himself and latterly of William Blake. Cosmological discoveries bring with them a feeling of awe but also, for most people, a sense of human diminishment. Our great sidereal adventures today are both elevating and frightening, and may be both at the same time. The launching of a space rocket is (to go back to seventeenth-century language) a tremendous phenomenon. It must have occurred to many who saw pictures of it that the great steel rampart or nave from which the Apollo rockets were launched had the size and shape and grandeur of a cathedral, with Apollo itself in the position of a spire. Like a cathedral it is economically pointless, a shocking waste of public money; but like a cathedral it is also a symbol of aspiration towards higher things.

When we compare the climates of opinion in the seventeenth century and today, we must again remember that cries of despair are not necessarily authentic. There was a strong element of affectation about Jacobean melancholy, and so there is today. Then as now it had tended to become a posture. One of a modern writer's claims to be taken seriously is to castigate complacency and to show up contentment for the shallow and insipid thing that it is assumed to be. But ordinary human beings continue to be vulgarly high spirited. The character we all love best in Boswell is Johnson's old college companion, Mr Oliver Edwards—the man who said that he had tried in his time to be a philosopher, but had failed because cheerfulness was always breaking in.

3

I should now like to describe the new style of thinking that led to a great revival of spirits in the seventeenth century. It is closely associated with the birth of science, of course—of Science with a capital S—and the 'new philosophy' that had been spoken of since the beginning of the century referred to the beginnings of physical science; but (as I said a little earlier) we should be taking too narrow a view of things if we supposed that the instauration of science made up the whole or even the greater part of it. The new spirit is to be thought of not as scientific, but as something conducive to science; as a movement within which scientific enquiry played a necessary and proper part.

What then were the philosophic elements of the new revival (using 'philosophy' again in its homely sense)?

The seventeenth century was an age of Utopias, though Thomas More's own Utopia was already years old. The Utopias or anti-Utopias we devise today are usually set in the future, partly because the world's surface is either tenanted or known to be empty, partly because we need and assume we have time for the fulfilment of our designs. The old Utopias—Utopia itself, the New Atlantis, Christianopolis, and the City of the Sun[9]—were contemporary societies. Navigators and explorers came upon them accidentally in far-off seas. What is the meaning of the difference? One reason, of course, is that the world then still had room for undiscovered principalities, and geographical exploration itself had the symbolic significance we now associate with the great adventures of modern science. Indeed, now that outer space is coming to be our playground, we may again dream of finding ready-made Utopias out there. But this is not the most important reason. The old Utopias were not set in the future because very few people believed that there would *be* a future—an earthly future, I mean; nor was it by any means assumed that the playing-out of earthly time would improve us or increase our capabilities.

On the contrary, time was running out, in fulfilment of the great Judaic tradition, and we ourselves were running down.

These thoughts suffuse the philosophic speculation of the seventeenth century until quite near its end. 'I was borne in the last Age of the World,' said John Donne,[10] and Thomas Browne speaks of himself as one whose generation was 'ordained in this setting of time'.[11] The most convincing evidence of the seriousness of this belief is to be found not in familiar literary tags, but in the dull and voluminous writings of those who, like George Hakewill,[12] repudiated the idea of human deterioration and the legend of a golden age, but had no doubt at all about the imminence of the world's end. The apocalyptic forecast was, of course, a source of strength and consolation to those who had no high ambitions for life on earth. The precise form the end of history would take had long been controversial—the New Jerusalem might be founded upon the earth itself or be inaugurated in the souls of men in heaven—but that history would come to an end had hardly been in question. Towards the end of the sixteenth century there had been some uneasy discussion of the idea that the material world might be eternal, but the thought had been a disturbing one, and had been satisfactorily explained away.[13]

During the seventeenth century this attitude changes. The idea of an end of history is incompatible with a new feeling about the great things human beings might achieve through their own ingenuity and exertions. The idea therefore drops quietly out of the common consciousness. It is not refuted, but merely fades away. It is true that the idea of human deterioration was expressly refuted—in England by George Hakewill but before him by Jean Bodin (by whom Hakewill was greatly influenced) and by Louis le Roy.[14] The refutation of the idea of decay did not carry with it an acceptance of the idea of progress, or anyhow of linear progress: it was a question of recognizing that civilizations or cultures had their ups and downs, and went through a life cycle of degeneration and regeneration—a 'circular kind of progress', Hakewill said.

There were, however, two elements of seventeenth-century thought that imply the idea of progress even if it is not explicitly affirmed. The first was the recognition that the tempo of invention and innovation was speeding up, that the flux of history was becoming denser. In *The City of the Sun* Campanella tells us that 'his age has in it more history within a hundred years than all the world had in four thousand years before it'. He is echoing Peter Ramus:[15] 'We have seen in the space of one age a more plentiful crop of learned men and works than our predecessors saw in the previous fourteen.' By the latter half of the seventeenth century the new concept had sunk in.

The second element in the concept of futurity—in the idea that men might look forward, not only backwards or upwards—is to be found in the breathtaking thought that there was no apparent limit to human inventiveness and ingenuity. It was the notion of a perpetual *plus ultra*, that what was already known was only a tiny fraction of what remained to be discovered, so that there would always be more beyond. Bacon published his *Novum organum* at the beginning of the remarkable decade between 1620 and 1630, and had singled it out as the greatest obstacle to the growth of understanding, that 'men despair and think things impossible'. 'The human understanding is unquiet', he wrote; 'it cannot stop or rest and still presses onwards, but in vain'—in vain, because our spirits are oppressed by 'the obscurity of nature, the shortness of life, the deceitfulness of the senses, the infirmity of judgement, the difficulty of experiment, and the like'. 'I am now therefore to speak of hope', he goes on to say, in a passage that sounds like the trumpet calls in *Fidelio*. The hope he held out was of a rebirth of learning, and with it the realization that if men would only concentrate and direct their faculties, 'there is no difficulty that might not be overcome'. '[T]he process of Art is indefinite,' wrote Henry Power, 'and who can set a *non-ultra* to her endeavours?'[16] There is a mood of exultation and glory about this new belief in human capability and the future in which it might unfold. With Thomas Hobbes 'glorying' becomes almost a technical term: 'Joy, arising from imagination of a man's own power and ability, is that exultation of mind called glorying', he says, in *Leviathan*, and in another passage he speaks of a 'perseverance of delight in the continual and indefatigable generation of knowledge'.

It does not take a specially refined sensibility to see how exciting and exhilarating these new notions must have been. During the eighteenth century, of course, everybody sobers up. The idea of progress is taken for granted—but in some sense it gets out of hand, for not only will human inventions improve without limit, but so also (it is argued, though not very clearly) will human beings. It is interesting to compare the exhilaration of the seventeenth century with, say, William Godwin's magisterial tone of voice as the eighteenth century draws to an end. 'The extent of our progress in the cultivation of human knowledge is unlimited. Hence it follows . . . [t]hat human inventions . . . are susceptible of perpetual improvement.'

Can we arrest the progress of the inquiring mind? If we can, it must be by the most unmitigated despotism. Intellect has a perpetual tendency to proceed. It cannot be held back, but by a power that counteracts its genuine tendency, through every moment of its existence. Tyrannical and sanguinary must be the measures employed for this purpose. Miserable and disgustful must be the scene they produce.[17]

The seventeenth century had begun with the assumption that a powerful force would be needed to put the inventive faculty into motion; by the end of the eighteenth century it is assumed that only the application of an equally powerful force could possibly slow it down.

Before going on, it is worth asking if this conception is still acceptable—that the growth of knowledge and know-how has no intrinsic limit. We have now grown used to the idea that most ordinary or natural growth processes (the growth of organisms or populations of organisms or, for example, of cities) is not merely limited, but self-limited, that is, is slowed down and eventually brought to a standstill *as a consequence of the act of growth itself.* For one reason or another, but always for some reason, organisms cannot grow indefinitely, just as beyond a certain level of size or density a population defeats its own capacity for further growth. May not the body of knowledge also become unmanageably large,

or reach such a degree of complexity that it is beyond the comprehension of the human brain? To both these questions I think the
answer is 'No'. The proliferation of recorded knowledge and the
seizing-up of communications pose technological problems for
which technical solutions can and are being found. As to the idea
that knowledge may transcend the power of the human brain: in
a sense it has long done so. No one can 'understand' a radio-set or
automobile in the sense of having an effective grasp of more than
a fraction of the hundred technologies that enter into their manufacture. But we must not forget the additiveness of human capabilities. We work through consortia of intelligences, past as well as
present. We might, of course, blow ourselves up or devise an
unconditionally lethal virus, but we don't *have* to. Nothing of the
kind is necessarily entailed by the growth of knowledge and
understanding. I do not believe that there is any intrinsic limita-
tion upon our ability to answer the questions that belong to the
domain of natural knowledge and fall therefore within the agenda
of scientific enquiry.

4

The repudiation of the concept of decay, the beginnings of a sense
of the future, an affirmation of the dignity and worthiness of
secular learning, the idea that human capabilities might have no
upper limit, an exultant recognition of the capabilities of man—
these were the seventeenth century's antidote to despondency.
You may wonder why I have said nothing about the promulgation
of the experimental method in science as one of the decisive
intellectual movements of the day. My defence is that the origin
of the experimental method has been the subject of a traditional
misunderstanding, the effect of reading into the older usages of
'experiment' the very professional meaning we attach to that
word today. Bacon is best described as an advocate of the *experi-
ential* method in science—of the belief that natural knowledge
was to be acquired not from authority, however venerable, nor by
syllogistic exercises, however subtle, but by paying attention to
the evidence of the senses, evidence from which (he believed) all

deception and illusion could be stripped away. Bacon's writings form one of the roots of the English tradition of philosophic empiricism, of which the greatest spokesman was John Locke. The unique contribution of science to empirical thought lay in the idea that experience could be *stretched* in such a way as to make nature yield up information which we should otherwise have been unaware of. The word 'experiment' as it was used until the nineteenth century stood for the concept of stretched or deliberately contrived experience; for the belief that we might make nature perform according to a scenario of our own choosing instead of merely watching her own artless improvisations. An 'experiment' today is not something that merely enlarges our sensory experience. It is a critical operation of some kind that discriminates between hypotheses and therefore gives a specific direction to the flow of thought. Bacon's championship of the idea of experimentation was part of a greater intellectual movement which had a special manifestation in science without being distinctively scientific. His reputation should not, and fortunately need not, rest on his being the founder of the 'experimental method' in the modern sense.[18]

Let us return to the contemporary world and discuss our misgivings about the way things are going now. No one need suppose that our present philosophic situation is unique in its character and gravity. It was partly to dispel such an illusion that I have been moving back and forth between the seventeenth century and the present day. Moods of complacency and discontent have succeeded each other during the past 400 or 500 years of European history, and our present mood of self-questioning does not represent a new and startled awareness that civilization is coming to an end. On the contrary, the existence of these doubts is probably our best assurance that civilization will continue.

Many of the ingredients of the seventeenth-century antidote to melancholy have lost their power to bring peace of mind today, and have become a source of anxiety in themselves. Consider the tempo of innovation. In the post-Renaissance world the feeling that inventiveness was increasing and that the whole world was on the move did much to dispel the myth of deterioration and give people confidence in human capability. Nevertheless the tempo was a pretty slow one, and technical innovation had little influence on the character of common life. A man grew up and grew old in what was still essentially the world of his childhood; it had been his father's world and it would be his children's too. Today the world changes so quickly that in growing up we take leave not just of youth but of the world we were young in. I suppose we all realize the degree to which fear and resentment of what is new is really a lament for the memories of our childhood. Dear old steam trains, we say to ourselves, but nasty diesel engines; trusty old telegraph poles but horrid pylons. Telegraph poles, as a Poet Laureate told us a good many years ago,[19] are something of a test case. Anyone who has spent part of his childhood in the countryside can remember looking up through the telegraph wires at a clouded sky and discerning the revolution of the world, or will have listened, ear to post, to the murmur of interminable conversations. For some people even the smell of telegraph poles is nostalgic, though creosote has a pretty technological smell. Telegraph poles have been assimilated into the common consciousness, and one day pylons will be, too. When the pylons are dismantled and the cables finally go underground, people will think again of those majestic catenary curves, and remind each other of how steel giants once marched across the countryside in dead silence and in single file. (What is wrong with pylons is that most of them are ugly. If only the energy spent in denouncing them had been directed towards improving their appearance, they could have been made as beautiful, even as majestic, as towers or bridges are allowed to be, and need not have looked incongruous in the countryside.)

When Bacon described himself as a trumpeter of the new philosophy, the message he proclaimed was of the virtue and dignity of scientific learning and of its power to make the world a better place to live in. I am continually surprised by the superficiality of the reasons which have led people to question those beliefs today. Many different elements enter into the movement to depreciate the services to mankind of science and technology. I have just mentioned one of them, the tempo of innovation when measured against the span of life. We wring our hands over the miscarriages of technology and take its benefactions for granted. We are dismayed by air pollution but not proportionately cheered up by, say, the virtual abolition of poliomyelitis. (Nearly 5,000 cases of poliomyelitis were recorded in England and Wales in 1957. In 1967 there were less than thirty.) There is a tendency, even a perverse willingness, to suppose that the despoliation sometimes produced by technology is an inevitable and irremediable process, a trampling down of nature by the big machine. Of course it is nothing of the kind. The deterioration of the environment produced by technology is a technological problem for which technology has found, is finding, and will continue to find solutions. There is, of course, a sense in which science and technology can be arraigned for devising new instruments of warfare, but another and more important sense in which it is the height of folly to blame the weapon for the crime. I would rather put it this way: in the management of our affairs we have too often been bad workmen, and like all bad workmen we blame our tools. I am all in favour of a vigorously critical attitude towards technological innovation: we should scrutinize all attempts to improve our condition and make sure that they do not in reality do us harm; but there is all the difference in the world between informed and energetic criticism and a drooping despondency that offers no remedy for the abuses it bewails.

Superimposed on all particular causes of complaint is a more general cause of dissatisfaction. Bacon's belief in the cultivation of science for the 'merit and emolument of life' has always been repugnant to those who have taken it for granted that comfort and prosperity imply spiritual impoverishment. But the real trouble nowadays has very little to do with material prosperity or technology or with our misgivings about the power of research and learning generally to make the world a better place. The real trouble is our acute sense of human failure and mismanagement, a new and specially oppressive sense of the inadequacy of men. So much was hoped of us, particularly in the eighteenth century. We were going to improve, weren't we?—and for some reason which was never made clear to us we were going to grow in moral stature as well as in general capability. Our school reports were going to get better term by term. Unfortunately they haven't done so. Every folly, every enormity that we look back on with repugnance can find its equivalent in contemporary life. Once again our intellectuals have failed us; there is a general air of misanthropy and self-contempt, of protest, but not of affirmation. There is a peculiar selfishness about modern philosophic speculation (using 'philosophy' here again in its homely or domestic sense). The philosophic universe has contracted into a neighbourhood, a suburbia of personal relationships. It is as if the classical formula of self-interest, 'I'm all right, Jack', was seeking a new context in our private, inner world.

We can obviously do better than this, and there is just one consideration that might help to take the sting out of our self-reproaches. In the melancholy reflections of the post-Renaissance era it was taken for granted that the poor old world was superannuated, that history had all but run its course and was soon coming to an end. The brave spirits who inaugurated the new science dared to believe that it was *not* too late to be ambitious, but now we must try to understand that it is a bit too early to expect our grander ambitions to be fulfilled. Today we are conscious that human history is only just beginning. There has always been room for improvement; now we know that there is time for improvement, too. For all their intelligence and dexterity—qualities we have always attached great importance to—the higher primates (monkeys, apes and men) have not been very successful. Human beings have a history of more than 500,000 years. Only during the past 5,000 years or thereabouts have human beings won a reward for their special capabilities; only during the past 500 years or so have they begun to be, in the biological sense, a success. If we imagine the evolution of living organisms compressed into one year of cosmic time, then the evolution of man has occupied a day. Only during the past ten to fifteen minutes of the human day has our life on earth been anything but precarious. Until then we might have gone under altogether or, more likely, have survived as a biological curiosity; as a patchwork of local communities only just holding their own in a bewildering and hostile world. Only during this past fifteen minutes (for reasons I shall not go into, though I think they can be technically explained) has there been progress, though, of course, it doesn't amount to very much. We cannot point to a single definitive solution of any one of the problems that confront us—political, economic, social or moral, that is, having to do with the conduct of life. We are still beginners, and for that reason may hope to improve. To deride the hope of progress is the ultimate fatuity, the last word in poverty of spirit and meanness of mind. There is no need to be dismayed by the fact that we cannot yet envisage a definitive solution of our problems, a resting-place beyond which we need not try to go. Because he likened life to a race,[20] and defined felicity as the state of mind of those in the front of it, Thomas Hobbes has always been thought of as the arch material-ist, the first man to uphold go-getting as a creed. But that is a travesty of Hobbes's opinion. He was a go-getter in a sense, but it was the going, not the getting, he extolled. As Hobbes conceived it, the race had no finishing post. The great thing about the race was to be in it, to be a contestant in the attempt to make the world a better place, and it was a spiritual death he had in mind when he said that to forsake the course is to die. 'There is no such thing as perpetual tranquillity of mind while we live here', he told us in *Leviathan*, 'because life itself is but motion and can never be without desire, or without fear, no more than without sense'; 'there can be no contentment but in proceeding'. I agree.

11 Further comments on psychoanalysis

In my Romanes Lecture on science and literature I implied that a psychoanalytical explanation-structure answered pretty closely to Lévi-Strauss's description of a myth. By this I meant that a psychoanalytical interpretation weaves around the patient a well-tailored personal myth within the plot of which the subject's thoughts and behaviour seem only natural, and, indeed, only what is to be expected.

I must begin by making it clear that my criticism of psychoanalysis is not to be construed as a criticism of psychiatry or psychological medicine as a whole. People nowadays tend to use 'psychoanalysis' to stand for all forms of psychotherapy, much as 'Hoover' is used as a generic name for all vacuum cleaners and 'Vaseline' for all ointments of a similar kind. By psychoanalysis I understand that special pedigree of psychological doctrine and treatment which can be traced back, directly or indirectly, to the writings and work of Sigmund Freud. The position of psychological medicine today is in some ways analogous to that of physical or conventional medicine in the middle of the nineteenth century. The physician of a hundred and thirty years ago was confronted by all manner of medical distress. He studied and tried to cure his patients with great human sympathy and understanding and with highly developed clinical skills, by which I mean that he had developed to a specially high degree that form of heightened sensibility which made it possible for him to read a meaning into tiny clinical signals which a layman or a beginner would have passed over or misunderstood. The physician's relationship to his patient was a very personal one, as if healing were not so much a matter of applying treatment to a 'case' as a collaboration between the physician's guidance and his patient's willingness to respond to it. But—there was so little he could do! The microbial theory of infectious disease had not been formulated, viruses were not recognized, hormones were unheard of, vitamins undefined, physiology was rudimentary and biochemistry almost non-existent.

The psychiatry of today is in a rather similar position, because we are still so very ignorant of the mind. But the best of its practitioners are people of great skill and understanding and apparently inexhaustible patience; people whose humanity reveals itself just as much in the way they recognize their limitations as in their satisfaction when a patient gets better in their care. I am emphasizing this point to make it clear that to express dissatisfaction with psychoanalysis is not to disparage psychological medicine as a whole.

One of my critics has accused me of saying or implying that he, a psychoanalyst, would attempt to treat by psychiatric means the symptoms of a brain tumour or of Huntington's Chorea. *Of course* I don't think a psychoanalyst would knowingly attempt to treat a brain tumour or a victim of Huntington's Chorea by psychoanalytic methods, but he may not realize the degree to which he is being wise after the event. Being a sensible man he naturally repudiates the idea of treating those psychological ailments of which physical causes are, in general terms, already known. But psychoanalysts do treat and speculate upon the origins of schizophrenic conditions and manic-depressive psychoses. *These* are the test cases: what are we to make of *them*?

Are 'mental illnesses' of mental or physical origin? To answer this question I shall begin with what may appear to be a digression. As recently as thirty years ago, many geneticists were still worried and confused by the problem of assessing, in precise terms, the relative contributions of nature and nurture—of heredity and environment or upbringing—to the overt ('phenotypic') differences between our mental and physical constitutions and capabilities. Both nature and nurture exercise an influence, of course; but L. T. Hogben and J. B. S. Haldane were the first to make it publicly clear that there is no *general* solution of the problem of estimating the size of the contribution made by each. The reason is that the size of the contribution made by nature is itself a function of nurture. (I use the word 'function' in its mathematical sense.)[1] If someone constitutionally lacks the ability to synthesize an essential dietary substance, say X, then the contribution made by heredity to the difference between himself and his fellow men will depend on the environment in which they live. If X is abundant in the food he normally has access to, his inborn disability will put him at no disadvantage and may not be recognized at all; but if X is in short supply or lacking, then he will become ill or die. The same reasoning applies to other, much more complicated examples. If people live a simple pastoral life that makes little demand on their resourcefulness and ingenuity, inherited differences of intellectual capability may not make much difference to their behaviour; but it is far otherwise if they live a difficult and intellectually demanding life. How often has it not been said that the stress of modern living raises the threshold of competence below which people can no longer keep up or make the grade? This is not to deny that some differences between us are for all practical purposes wholly genetic, wholly inborn. A person's blood group is described as 'inborn' not just because it is specified by his genetic make-up, but because (with certain rare and known exceptions) there is no environment capable of supporting life in which that specification will not be carried out. Most differences between us are determined both by nature and by nurture, and their contributions are not fixed, but vary in dependence on each other.[2]

With this analogy in mind, let me now turn to psychological disorders, which—to beg no questions—I shall define as conditions which cause a person to seek, or need, or be directed towards, the care of a psychiatrist. Here, too, as a first approximation, it will be reasonable to assume that both 'mental' and 'organic' states or agencies contribute to the difference between the psychiatrist's patient and his fellow men; but here, too, we should be very cautious in our attempts to assign precise values to the contributions made by each. It seems natural to repudiate the idea of psychiatric treatment of brain tumours, because they seem so obviously organic in origin; but even in this extreme case we mustn't be too sure. Many of us now believe that there exists a natural defensive mechanism against tumours which is of the same general kind as that which prohibits the transplantation of tissues from one individual to another. If these natural defences are indeed immunological in nature, they are open to influences of a kind that common sense will classify as mental, or anyhow behavioural, e.g. to prolonged frustration, unhappiness, distress, or indifference to living. (The psychosomatic element in tuberculosis is specially relevant here, because the natural defence against tuberculosis depends on immunological mechanisms of a very similar kind.)

To go now to the other extreme: the psychoanalytic critic I referred to above thinks it probable that 'neurosis is the result of faulty early conditioning' rather than of brain disease or an inborn error of metabolism. No doubt; but does he not also think that constitutional or organic influences may raise or lower the susceptibility of his patients to these disturbing influences? Of course he does—and so did Freud. It is normally a mistake, I suggest, to trace any psychological disorder to wholly mental or wholly organic causes. Both contribute, though sometimes to very unequal degrees, and the contribution made by one will be a function of the contribution made by the other.

It is, nevertheless, very understandable that psychiatrists should approach their patients with two rather different kinds of aetiological purpose and interest in mind. Psychiatrist A will say, 'My interest lies in trying to see how a certain pattern of upbringing, environment, habits of life and human relationships may predispose people of certain constitutions to psychological disorders.' Psychiatrist B will say, 'Now *my* interest lies in trying to identify those elements of heredity and organic constitution which make a man specially likely to contract a certain psychological disorder if he is influenced by the environment and his fellow men in certain ways.' Both attitudes seem very reasonable, and over much of the territory that belongs to them the two psychiatrists will not compete. But—and now I come to my main point—in the context of those serious psychological disorders that are still disputed territory, the methodology implicit in the attitude of Psychiatrist B is very much the more powerful.

The reason is this. A physical abnormality can be the subject of diagnosis, and therefore, in principle, of treatment or avoiding action, *before* it can contribute to a psychological disturbance. The recognition early in life of a certain physical abnormality (say, the chromosomal constitution XYY) defines a priori a category of men who are at special risk; and our foreknowledge of that risk can be made the basis of a rational system of avoidance. The physical disability represents a parameter of the situation, where upbringing and environment are variables which can be varied within certain limits at our discretion. A difficult enterprise, to be sure; but not so difficult as, and much more realistic than, say, abolishing all family life, as one 'existential psychiatrist' is alleged to have recommended, because some families may create an environment conducive to mental disorder. With certain forms of low-grade mental deficiency, this programme is now adopted as a matter of routine. When tests carried out on a baby's urine suggest that it cannot metabolise the amino-acid phenylalanine, its diet can be altered in such a way as to prevent or reduce the severity of what might otherwise be irremediable damage to the brain. I hope and expect that cognate solutions will one day be found for the major psychoses. No matter what other factors may have influenced him, there is something organically wrong with a manic-depressive patient, and it is essential to find out what it is, preferably before he becomes gravely ill.

This completes my attempt to explain why I think that the categorical distinction between brain disease and mental illness, as between 'Nature' and 'Nurture', is a fundamentally unsound one—the remnant of an effete dualism, a still further perpetuation of what Ryle called the legend of Two Worlds.

I now turn to psychoanalysis itself, taken in the sense I gave it in an earlier paragraph. I shall not attempt a systematic treatment, but shall merely draw attention to a few of its more serious methodological, doctrinal and practical defects.

The property that gives psychoanalysis the character of a mythology is its combination of conceptual barrenness with an enormous facility in explanation. To criticize a theory because it explains everything it is called on to explain sounds paradoxical, but anyone who thinks so should consult the discussion by Karl Popper in *Conjectures and Refutations*, particularly the passages that make mention of psychoanalysis itself.[3] Let me illustrate the point by a number of passages chosen from the authors' summaries of their own contributions to be 23rd International Psychoanalytical Congress held in Stockholm in 1963. I choose the proceedings of a congress rather than the work of a single author so as to get a cross-section of psychoanalytic thought.

Character-traits are formed as precipitates of mental processes. They originate in innate properties; they come into existence in the mutual interplay of ego, id, super-ego and ego-ideal, under the influence of object-relations and environment.

When an individual strikes out at his wife, his child, his acquaintances or even complete strangers, we may well suspect that a gross failure in Ego-functioning has occurred. Its restraining control has been partially eluded.

Of a 'cyclothymic' patient in the fifth and sixth years of psychoanalytic treatment:

. . . the delusion of having black and frightening eyes took the centre of the analytic stage following the resolution of some of the patient's oral-sadistic conflicts. It proved to be a symptom of voyeuristic tendencies in a split-off masculine infantile part of the self and yielded slowly to reintegration of this part, passing through phases of staring, looking at and admiring the beauty of women.

On the aetiology of anti-Semitism:

The Oedipus complex is acted out and experienced by the anti-Semite as a narcissistic injury, and he projects this injury upon the Jew who is made to play the role of the father . . . His choice of the Jew is determined by the fact that the Jew is in the unique position of representing at the same time the all-powerful father and the father castrated . . .

On the role of snakes in the dreams and fantasies of a sufferer from ulcerative colitis:[4]

The snake represented the powerful and dangerous (strangling), poisonous (impregnating) penis of his father and his own (in its anal-sadistic aspects). At the same time, it represented the destructive, devouring vagina . . . The snake also represented the patient himself in both aspects as the male and female and served as a substitute for people of both sexes. On the oral and anal levels the snake represented the patient as a digesting (pregnant) gut with a devouring mouth and expelling anus . . .

I have not chosen these examples to poke fun at them, ridiculous though I believe them to be, but simply to illustrate the Olympian glibness of psychoanalytic thought. The contributors to this congress were concerned with homosexuality, anti-Semitism, depression, and manic and schizoid tendencies; with *difficult* problems, then—problems far less easy to grapple with or make sense of than anything that confronts us in the laboratory. But where shall we find the evidence of hesitancy or bewilderment, the avowals of sheer ignorance, the sense of groping and incompleteness that is commonplace in an international congress of, say, physiologists or biochemists? A lava-flow of *ad hoc* explanation pours over and around all difficulties, leaving only a few smoothly rounded prominences to mark where they might have lain. Surely the application of psychoanalytic methods in a completely alien culture might give even the most sanguine practitioner reason to pause? Not a bit of it. We have the word of two of the contributors to the congress that 'the usual technique and theory of psychoanalysis were found to be applicable to obtain an understanding of the inner life' of the Dogon peoples in Mali:

A twenty-four-year-old Dogon man, who at the beginning had met the white stranger with profound distrust, was led to change his views with surprising speed.

After first having built a subsidiary transference and involved a younger colleague in the analysis, he turned from the animate object to the inanimate (playing with sticks) and from this to tactile gestures . . . Finally he 'regressed' to somatic forms of expression in that he continued the analytic exchange by urinating . . .

The examples I have chosen above, and the psychoanalytic autopsies I shall mention later, illustrate another important methodological defect of psychoanalytic theory. If an explanation or interpretation of a phenomenon or state of affairs is to be fully satisfying and actable-on, it must have a special, not merely a general relevance to the problem under investigation. It must be, rather specially, an explanation of whatever it is we want to explain, and not also an explanation of a great many other, perhaps irrelevant things as well.

For example: if a patient cannot retain salt in his body, it is not good enough (though it will probably not be wrong) to say that his endocrine system is in disorder, because such an explanation would cover a multitude of other abnormalities besides. The explanation may well be that the patient is no longer producing aldosterone, a specific hormone of the cortex of the adrenal gland, and if that is so he can probably be cured. Again, it will not do to say that muscular contraction is a transformation of energy derived originally from the sun. This is a weak explanation; it is too far removed in the pedigree of causes; we are more interested in the causal parentage of the phenomenon than in its causal ancestry. Strong explanations have a quality of *special* relevance, of logical immediacy: and this is a quality they must have if they are to be tested and shown to be acceptable for the time being or, as the case may be, unsound. Psychoanalytic explanations are invariably weak explanations in just this sense.

'Validation of psychoanalytic theory is a difficult business', my psychoanalytic disputant said, though he betrayed no logical understanding of why it should be so; and by implication he suggested that, instead of criticizing it destructively, I should help find means of testing whether or not it is true. Alas—except in one respect, which I shall deal with in a moment—the methodological obstacles are insuperable. Indeed, psychoanalysis has now achieved a complete intellectual closure: it explains even why some people disbelieve in it. But this accomplishment is self-defeating, for in explaining why some people do not believe in it, it has deprived itself of the power to explain why other people do. The ideas of psychoanalysis cannot both be an object of critical scrutiny and at the same time provide the conceptual background of the method by which that scrutiny is carried out.

It is for this reason that the notion of *cure* is methodologically so important. It provides the only independent criterion by which the acceptability of psychoanalytic notions can be judged. This is why cure is such an embarrassment for 'cultural' psychiatry in general. No wonder its practitioners try to talk us out of it,[5] no wonder they prefer to see themselves as the agents of some altogether more genteel ambition, for example, to give the patient a new insight through a new deep, inner understanding of himself. But let us not be put off. Some people get better *under* psychoanalytic treatment, of course; but do they get better as a specific consequence of psychoanalysis as such? Consider an example. A young man full of anxieties and worries seeks treatment from a psychoanalyst, and after eighteen months' or two years' treatment finds himself much improved. Was psychoanalytic treatment responsible for the cure? One cannot give a confident answer unless one has reasonable grounds for thinking:

(*a*) that the patient would not have got better anyway;

(*b*) that a treatment based on quite different or even incompatible theoretical principles, for example, the theories of a rival school of psychotherapists, would not have been equally effective; and

(*c*) that the cure was not a by-product of the treatment. The assurance of a regular sympathetic hearing, the feeling that somebody is taking his condition seriously, the discovery that others are in the same predicament, the comfort of learning that his condition is explicable (which does not depend on the explanation's being the right one)—these factors are common to most forms of psychological treatment, and the good they do must not be credited to any one of them in particular. At present there is no convincing evidence that psychoanalytic treatment as such is efficacious, and unless strenuous efforts are made to seek it the entire scheme of treatment will degenerate into a therapeutic pastime for an age of leisure.

The lack of good evidence of the specific therapeutic effectiveness of psychoanalysis is one of the reasons why it has not been received into the general body of medical practice. A layman might be inclined to say that we should give it time, for doctors are conservative people and ideas so new take ages to sink in. But it is only on a literary time-scale that Freudian ideas are new. By the standards of current medical practice they have an almost antiquarian flavour. Many of Freud's principles were formulated before the recognition of inborn errors of metabolism, before the chromosomal theory of inheritance, before even the rediscovery of Mendel's laws. Hormones were unheard of when Freud began to propound his doctrines, and the mechanism of the nervous impulse, of which we now have a pretty complete understanding, was quite unknown.

Nevertheless, psychoanalysts are wont to say that Freud's work carried conviction because it was so firmly grounded on basic biological principles. I am therefore sorry to have to express the professional opinion that many of the germinal ideas of psycho-analysis are profoundly unbiological, among them the 'death-wish', the underlying assumption of an extreme fragility of the mind, the systematic depreciation of the genetic contribution to human diversity, and the interpretation of dreams as 'one member of a class of *abnormal* psychical phenomena'.

I said earlier that the mythological status of psychoanalytic theory revealed itself in its combination of unbridled explanatory facility with conceptual barrenness, a property to which I have not yet referred. Ever since Freud's factually erroneous analysis of Leonardo,[6] psychoanalysts have tried their hand at 'interpreting' the life and work of men of genius, and many of the great figures of history have been disinterred and brought to the post-mortem slab. The fiasco of Darwin's retrospective psychoanalysis has already been held up to ridicule.[7] But, Darwin apart, how can we not marvel at the way in which the whole exuberant variety of human genius can be explained by the manipulation of a handful of germinal ideas—the Oedipus complex, the puzzlement of dis-covering that not everyone has a penis, a few unspecified sado-masochistic reveries, and so on: surely we need a more powerful armoury than this? Evidently we do, for these analyses always stop short of explaining why genius took the specific form that interests us. Freud does not profess to tell us why Leonardo became an artist. 'Just here our capacities fail us', he says, with a modesty not found in the writings of his successors; but it is hard not to feel let down.

A critique of psychoanalysis is, in the outcome, never much more than a skirmish, because (as I tried to explain) its doctrines are so cunningly insulated from the salutary rigours of disbelief. It is nevertheless customary to end any such critique with a spaciously worded acknowledgement of our indebtedness to Freud himself. We recognize his enlargement of the sensibilities of physicians, his having opened up a new era of human speculation, his freeing us from the confinements of prudery and self-righteousness, etc. There is some truth in all of this. There is some truth in psychoanalysis too, as there was in Mesmerism and in phrenology (for example, the concept of localization of function in the brain). But, considered in its entirety, psychoanalysis won't do. It is an end-product, moreover, like a dinosaur or a zeppelin; no better theory can ever be erected on its ruins, which will remain for ever one of the saddest and strangest of all landmarks in the history of twentieth-century thought.

Postscript

My methodological criticisms of psychoanalysis deserve more attention than any psychoanalyst has yet found time or inclination to give them. While reaffirming my belief that these criticisms are valid let me again emphasize that they are expressly directed against psychoanalysis and not against psychiatry in general. This methodological criticism is of course far from complete. Professor Hugh Trevor-Roper (Lord Dacre) in the *Sunday Times* of 18 February 1973 has called attention to another methodological enormity—one specially perpetrated by 'psycho-historians'. The usual practice in science or historical research is to frame hypotheses in such a form that the facts follow from them, that is, in such a way that statements expressing the matters of fact in need of explanation are among the logical implications of the hypothesis. In psycho-history, however, the facts are shaped in such a way as to make them appear to follow from a preconceived hypothesis.

This psycho-historical approach authorizes us to declare with certainty that Hitler's character make-up and behaviour point to Mrs Hitler's extreme severity with young Adolf's toilet-training, a subject of which we are luckily quite ignorant.

12 The strange case of the spotted mice

'Can the leopard change his spots?' the prophet asked (Jeremiah, 13:23), clearly not expecting to be told he can. Nor, indeed, can mice, except under the rather discreditable circumstances now to be outlined.

It is a well-attested truth of observation that except under special and unusual circumstances skin from one mouse or human being will not form a permanent graft after transplantation to another mouse or another human being; for although such a graft heals into place it soon becomes inflamed and ulcerated, and eventually dries up and sloughs off. The exceptional circumstances are: in human beings, when donor and recipient are identical twins, and in mice when prolonged inbreeding (e.g., upward of twenty successive generations of brother/sister mating) has made the mice so closely similar to each other genetically that they almost could be identical twins.

This being so, great surprise was caused in the world of transplantation when Dr William Summerlin, a member of the largest and in many ways the most important cancer research centre in the world, the Sloan-Kettering Institute in New York, with the backing of his chief, Dr Robert A. Good, made known in 1973 his surprising claim that a comparatively simple procedure—'tissue-culture'—could make a skin graft or a corneal graft from a member of the same or even of a different species acceptable to an organism that would otherwise have rejected it. This claim was specially important because grafting skin from one human being to another has never entered clinical practice, in spite of encouraging successes with the transplantation of kidneys, livers, and sometimes even hearts. Either skin is specially well able to excite the immunological reaction that leads to its own rejection, or it is specially vulnerable to it. This inability to graft new skin from one person to another is the greatest current shortcoming of the surgery of replacement and repair, because the replacement of skin is the only adequate treatment of extensive burns or excoriating wounds.

Summerlin's treatment, the technical details of which, in spite of exhortation from his director, he seemed suspiciously reluctant to impart to his colleagues, amounted in principle to very little more than the incubation of the intended graft in a suitable nutrient medium outside the body for a matter of days or weeks. This seemed an astonishingly simple solution of a problem no one else had solved, although many of us had been trying since about 1940.

Unfortunately, experienced biologists in other laboratories, and eventually workers in the same institute, were unable to confirm Summerlin's findings, so that Summerlin eventually had recourse to faking his results to convince his now uneasy chief. He touched up his grafts with a felt pen, so simulating dark skin grafts on white mice. He also claimed that operations had been done which had not been done. The formal end of the story came in 1974 when Dr John L. Ninnemann and Dr Good published a paper that in effect demolished the whole story.

This was all a nine days' wonder in the world of immunology, but the nine days are now up and this is therefore a good moment, on the basis of Hixson's very readable and, so far as I can tell, very accurate account[1] of the whole story, to stand back and see what lessons can be learned from the whole episode. This is also Hixson's ambition, for he says at the outset: 'I hope that by the time the reader has reached the end of the book, he or she will have enough information to form an opinion about what is good and what is not so good in our current system of medical research as it pertains to cancer.' 'If the reader disagrees with the author,' he goes on bravely, 'why then, so much the better.'

Summerlin's sin is not now in doubt; but it is still worth considering precisely why his action was considered so heinous by all his fellow scientists. The reason is this: scientists try to make sense of the world by devising hypotheses, i.e., draft explanations of what the world is like; they then examine these explanations as critically as they know how to, with the result that either they gain confidence in their beliefs or they modify or abandon them.

In the ordinary course of events scientists very often guess wrong, take a wrong view, or devise hypotheses that later turn out to be untenable. This is an ordinary part of human fallibility and calls for no special comment. Nor does it necessarily impede the growth of science because where they themselves guess wrong, others may yet guess right. But they won't guess right if the factual evidence that led to formulating the hypothesis and testing its correspondence with reality is not literally true. For this reason, any kind of falsification or fiddling with professedly factual results is rightly regarded as an unforgivable professional crime.

In trying unsuccessfully to get the same results as Summerlin, my colleagues and I wasted a lot of time that might have been much more fruitfully employed. Our failure—and the failure of others—to repeat his results was not in itself irremediably damaging, for this, too, is an ordinary part of scientific life. After Rupert Billingham, Leslie Brent, and I published experiments showing quite clearly that the problem of how to overcome the incompatibility barrier between unrelated individuals was indeed soluble, several people tried to repeat our work and failed. There were, however, always good reasons why they did so; either they had introduced into our techniques little 'improvements' of their own, or they were too clumsy or something. These failures did not disturb us in the very least: we knew we were right—and we were—so we did our best to tell those who were struggling with our techniques how best to carry them out. As Hixson makes plain, Summerlin was suspiciously at fault, for he simply would not divulge his methods. Indeed, matters reached such a point that Leslie Brent, one of the world's foremost experts on transplantation, was driven in desperation to send a whole file of his correspondence with Summerlin to Dr Good, an action unwillingly taken which led to Summerlin's being severely reprimanded.

A particularly exasperating characteristic of the whole episode was that Summerlin's claim *could* easily have been true and for reasons which Dr Good described as 'trivial'. They would have been trivial only because they did not point to any scientifically exciting phenomenon such as change of genetically programmed characteristics in the graft, for example its makeup of immunity-provoking substances. But from the point of view of clinical usefulness it obviously didn't matter whether the explanation was profound or trivial. So many of us—even those who like myself shared Good's view that the reasons for the grafts' anomalous 'take' were trivial—persevered in trying to repeat Summerlin's work.

I am desperately sorry that Summerlin's work turned out to be mistaken because its failure means that we are still without means of repairing the skin surface except by piecemeal patching with little fragments of the patient's own skin—a process that may take weeks or even months during which the patient steadily loses body fluids and is specially vulnerable to infection.

The reader may well want to know what the very distinguished members of the Board of Scientific Consultants of the Sloan-Kettering Institute were up to all this time. My name appears repeatedly in Hixson's book as an expert on transplantation and partly as a member of the board. I cut a better figure in the pages of Hixson's book than I did in real life—something for which I bear Hixson no ill will. My reason for saying so is that at several critical points I found myself lacking in moral courage.

Summerlin once demonstrated to our assembled board a rabbit which, he said, had received from a human being a 'limbus to limbus' corneal graft—a graft which had been made compatible by his process of culturing. 'Limbus to limbus' means extending over the whole dome of the cornea to the extreme rim in which the blood vessels run. Through a perfectly transparent eye this rabbit looked at the board with the candid and unwavering gaze of which only a rabbit with an absolutely clear conscience is capable.

I could not believe that this rabbit had received a graft of any kind, not so much because of the perfect transparency of the cornea as because the pattern of blood vessels in the rim around the cornea was in no way disturbed. Nevertheless, I simply lacked the moral courage to say at the time that I thought we were the victims of a hoax or confidence trick. It is easy in theory to say these things, but in practice very senior scientists do not like trampling on their juniors in public. Besides, it was still possible that for some reason, 'trivial' or otherwise, the story was true. However we made no secret of our inability to repeat some of Summerlin's experiments, so far as we were able on the basis of the very inadequate information we had.

On the one occasion when I visited Summerlin in his laboratory with his immediate coworkers and technical helpers, and asked a number of hostile questions, I noticed with some surprise that our duologue was causing the others quite a lot of amusement. In retrospect, and after learning from Hixson the part Summerlin's technicians and immediate coworkers played in showing up the counterfeit, I can now see that they were sardonically amused at Summerlin's being interrogated in this way. But the equally plausible hypothesis I formed at the time was that, being only human, they were in reality amused at the obvious discomfiture of an eminent visiting scientist who, from the nature of his position on the board, was 'one of *them*' rather than 'one of *us*'. But for whatever reason, I did not forthrightly express any grave doubts about the probity of the whole enterprise.

In cases such as Summerlin's it is the usual thing to go over the culprit's career to find premonitions in his early life of how he behaved later. Hixson has done a good job here, reporting upon an unproved charge that Summerlin cheated in exams during his sophomore year at Emory University School of Medicine. It is, of course, possible that Summerlin was what is known in the world of criminology as a 'bad apple', but this diagnosis lacks psychological depth.

I believe that there is a fairly simple explanation of Summerlin's egregious folly. It is this: in his early experiments Summerlin *did* actually obtain with mice, the results that later aroused so many misgivings. Mice can sometimes get muddled up even in the best regulated laboratories, and it is just conceivable that in his earliest experiments the recipient mice which Summerlin believed to be genuinely incompatible with their donors were in reality hybrids between the strain of the graft donor and some other mouse.

For genetic reasons, such hybrids would have accepted the donor's skin grafts anyway—irrespective of the 'tissue culture'; if this is what happened then Summerlin would naturally have been distraught when, on repeating his experiments with genuinely incompatible mice, he found they didn't work. Being absolutely convinced in his own mind that he was telling a true story, he thereupon resorted, disastrously, to deception. A mistake exactly analogous to this was once made by a securely established American expert on transplantation who reported an equally implausible result to the Transplantation Society; so far from losing face—except perhaps in his own mirror—this young man gained some credit for withdrawing his results and frankly admitting that he had made a booboo.

Every scientist is at all times aware of his own fallibility and of the special safeguards that must be taken to avoid biasing the interpretation of results in a way that favours some hypothesis he may be temporarily in love with. 'Leaning over backward' is a well-known formula for avoiding self-deception—it stands for making sure that errors of observation arising from uncontrollable sources always tell *against* the hypothesis we should like to see corroborated. It is for this reason, too, that clinical trials of new remedies have to be done 'double blind'. Neither the patients nor the clinicians must know which patients are receiving a new wonder drug and which a mere placebo. A disinterested third party holds the key and will not unlock the code until clinical assessments are complete, after which it may, unhappily, turn out that 250 mg per day of placebic acid is as good a preventative of common colds as ascorbic acid (vitamin C).

My interpretation of Summerlin's behaviour is not intended to depreciate the importance and incentive to a scientist of public recognition and the esteem of one's colleagues. Election into the Fellowship of the Royal Society or its equivalent elsewhere is an honour because it is a public recognition of the fact that one's colleagues admire and applaud one's work. What is still mysterious it that a man of Summerlin's obvious intelligence and ability could have supposed he would get away with it, unless indeed, as I have suggested, he *did* once get the results he ultimately faked and thus felt perfely confident that workers elsewhere would, in the fullness of time, uphold his claims.

How does Robert Good, the director of Summerlin's institute, come out of all this?

There is no more important position in medical science than the directorship of the Sloan-Kettering Institute, a position for which Dr Robert Good was qualified by being enormously knowledgeable, brilliantly clever, persuasively eloquent, and indefatigably hard-working. Such a man naturally accumulates enemies, many of whom must have felt several inches taller when the Summerlin affair threw discredit upon him. Quite the most sickening aspect of the whole business was the way in which so many people whose lives had, until then, been devoted single-mindedly to self-advancement sprang into moral postures, pursed their lips, and moralized in a vein of excruciating triteness and dullness.

The Summerlin affair, we were told, was the natural consequence of the prosecution of science in an abrasively competitive world with limited research funds. If Good attached such importance to the work and sponsored his colleague so eloquently in public, then why did he not take more pains to supervise the research and make sure that everything was as it professed to be?

The answers to these questions do not wholly exculpate Good (and he himself does not think they do), but they show him in a very much better light than his enemies would like. In the first place it is a very much more endearing trait in the head of an institute to champion and promote the interests of his young than to let them get on with it while he busies himself with his own affairs. It was, indeed, Good's patronage in Minneapolis that made it possible for Summerlin to have a career at all. In the second place it is not physically possible for the head of an institute of several hundred members to supervise intently the work of each one. Most such directors assume—and as a rule rightly—that young recruits from good schools and with good references will abide by the accepted and well-understood rules of professional behaviour. I write here with the authority of someone who has been the head of a large research institute, and who has himself been once or twice deceived by an impostor.

Summerlin's attitude toward Good is made clear by his state-ment that he went to the Sloan-Kettering because Good was there: 'In retrospect I have to plead guilty to an overdose of hero worship. You know, I felt very close to this man. Regrettably, it wasn't mutual, as it turned out.' If this last, mean-minded comment is characteristic of Summerlin, it makes much else intelligible.

An especially attractive feature of Hixson's book is the evidence of the way in which science writers collude with each other and with scientists to create an atmosphere that will help them raise funds for their research. Summerlin was evidently a beneficiary of this process and it shows great forbearance on Hixson's part that he is not more indignant than he is on behalf of his fraternity of science writers.

Hixson's writing is of the quality we have come rather compla-cently to expect nowadays from first-rate professional science writers, although there are occasional lapses. On the page Summerlin is described as a 'tall, balding young skin specialist', in the idiom *Time* magazine has accustomed us to. Would it have mattered if Summerlin had been stocky and bushy-haired? Per-haps, for elsewhere Hixson makes it pretty clear that Summerlin's charm, enthusiasm, and general plausibility were not the least important part of his overall strategy for self-advancement.

The help science writers can give is very important since nearly all biomedical research—and particularly cancer research—is enormously costly, and the passages in which Hixson describes how it is financed will particularly interest not only potential private benefactors but also those needy scholars, especially in the humane arts, who follow less richly endowed pursuits. By one means or another—whether by private benefactions or fiscal levies—it is the general public that finances all cancer research, and this is how it should be, for they are ultimately the beneficiaries.

By the standards prevailing in humanistic studies the sums available are very large, but they are not to be had merely for the asking: one or more—sometimes a tier—of expert committees, all notoriously hard to please, stand between the applicant and the moneybags. Administrative committees refer a grant application to the expert body most highly qualified to express an informed opinion upon it and sometimes to other experts interested in cognate research.

Service on the National Institutes of Health Study Sections in America and on various boards of the Research Councils in England is enormously laborious and time-consuming, particularly as it may sometimes, as with Summerlin's application, entail personal visits to the laboratories in which the applicants are working. But so well understood is the importance of the task these study boards perform that it is possible to recruit to them extremely busy and able young scientists, often at stages in their careers when they can ill afford to give up their time for the purpose. Among these young bloodhounds are often a number of seniors who, having themselves probably been beneficiaries, have seen it all before and can therefore help to prevent hasty or injudicious decisions.

The study group that visited Summerlin's laboratories evidently had some misgivings about the authenticity of the work, but however deep-seated these may have been Summerlin was funded both by the NIH and by other benefactors.[2] Does this point to something deeply wrong with the present system of research funding—to something which, for the protection of the public, should be remedied right away?

Speaking as a man who has been a beneficiary of NIH research funds and who has served many severe sentences on grant-giving bodies, I should say not. In cost benefit, I should say that the most successful grant-giving bodies and research sponsors in the world are the Max Planck Gesellschaft in Germany and the Medical and Agricultural Research Councils in Great Britain. They have made grievous blunders, of course—including blunders of the kind that grant-giving bodies most greatly dread, i.e., the failure to fund research that has ultimately turned out to be of the greatest importance. But in the main they do whatever can be expected of bodies of highly informed and concerned scientists. Fortunately, clairvoyance and mind-reading are beyond them. It is not logically possible to predict future theories or future ideas, or, therefore, to be sure that a particular theory proposed for investigation will yield a harvest of fruitful ideas that will stand up to determined self-criticism.

However, there is an ignorance—amounting sometimes almost to contempt—of scientific philosophy not only among scientists but also among people, professedly critical thinkers, who ought to know, and often profess to know, better. This has led to the widespread misconception that the scientist works according to the rules of some cut and dried intellectual formulary known as 'the scientific method'. It has therefore come to be widely believed that given money and resources a scientist can bend the scientific method to the solution of almost any problem that confronts him. If he does not, it can only be because he is lazy or incompetent. In real life it is not like that at all. It cannot be too widely understood that there is no such thing as a 'calculus of scientific discovery'. The generative act in scientific discovery is a creative act of mind—a process as mysterious and unpredictable in a scientific context as it is in any other exercise of creativity.

We cannot devise hypotheses to order. Shelley would have understood this perfectly, for in his *Defence of Poetry* he wrote: 'A man cannot say "I *will* compose poetry"; The greatest poet even cannot say it. . . .' Nor can even the greatest scientist undertake to have illuminating ideas upon any problem he is confronted with, though he will know probably from experience how to put himself in the right frame of mind for getting ideas, and what reading and discussions will help him have them.

For these reasons most grant-giving bodies have come empirically to understand that they are most likely to do good by supporting people rather than projects, though I myself think this is a confession of weakness, for while conceding that no committee can do research, I nevertheless think that a committee which really knows its business should be able not merely to formulate a problem but also to indicate the lines along which it is most likely to be solved. Even if it were wrong, as very likely it would be, its thinking on the matter might easily spark off some fruitful idea in the mind of the investigator it commissioned to undertake its project.

I do not, however, think there is anything radically wrong with our present grant-giving procedures or that the grant-giving agencies can be convicted of anything more serious than of sometimes making mistakes. The romantic view of the creative process of science as something cognate with poetic invention is often sneered at by people who pride themselves as shrewd, practical-minded men of the world with a sound sense of the value of money. But they don't do any better than the rest of us, and it is they, indeed—people who believe that there is a cut and dried scientific method and that they can buy scientific results by paying for them—who are the incurable daydreamers with their heads in the clouds and no real understanding of the way the mind works.

It is characteristic of Hixson's balanced and fair-minded account of the Summerlin affair, which is likely to be the definitive account of the whole business, that he cites criticisms of his own profession. His discussion of the relationship between scientists and the press during an annual meeting of the Federation of American Societies for Experimental Biology makes it obvious that science writers want clear stories without the cagey reservations scientists are always introducing, and this may do something to incite people of Summerlin's temperament—though goodness knows he needed little encouragement.

How do scientists come out of it all? When the Summerlin affair became known, laymen shook their heads regretfully and exchanged long, significant looks as if to imply that they had learned something profoundly new about the scientific life and the morals of scientists. This is because two stereotypes of scientists dominate the lay imagination: the first is a figure like Martin Arrowsmith with a chronically dedicated expression on his face who is willing to sacrifice wealth and an easy life, even love and personal advancement, to the discovery of the new serum upon which he is covertly working after his colleagues have left college or laboratory for the night. The second is a Gothic figure intent on devising ever more expeditious means of destroying the human race—a man who, as his work comes to fruition, cries out in a strong Central European accent (for no American or Britisher could be guilty of such behaviour), 'And soon ze whole vorld vill be in my power' (maniacal laughter). In reality there are all kinds of different people who are scientists. I once put the matter thus:[3]

Scientists are people of very dissimilar temperaments doing very different things in very different ways. Among scientists are collectors, classifiers and compulsive tidiers-up; many are detectives by temperament and many are explorers. Some are artists and others are artisans. There are poet-scientists and philosopher-scientists, and even a few mystics.

If only I had thought to add '. . . and just a few odd crooks', then I should have drawn a clear distinction between the scientific profession and the pursuit of mercantile business, politics, or the law, professions of which the practitioners are inflexibly upright all the time. As it is, I am afraid no great truth about scientific behaviour is to be learned from the Summerlin affair except perhaps that it takes all sorts to make a world.

13 Unnatural science

If a broad line of demarcation is drawn between the natural sciences and what can only be described as the unnatural sciences, it will at once be recognized as a distinguishing mark of the latter that their practitioners try most painstakingly to imitate what they believe—quite wrongly, alas for them—to be the distinctive manners and observances of the natural sciences. Among these are:

(*a*) the belief that measurement and numeration are intrinsically praiseworthy activities (the worship, indeed, of what Ernst Gombrich calls *idola quantitatis*);

(*b*) the whole discredited farrago of inductivism—especially the belief that facts are prior to ideas and that a sufficiently voluminous compilation of facts can be processed by a calculus of discovery in such a way as to yield general principles and natural-seeming laws;

(*c*) another distinguishing mark of unnatural scientists is their faith in the efficacy of statistical formulae, particularly when processed by a computer—the use of which is in itself interpreted as a mark of scientific manhood. There is no need to cause offence by specifying the unnatural sciences, for their practitioners will recognize themselves easily: the shoe belongs where it fits.

The objections of the educated to IQ psychology arise from several sets of causes: first, misgivings about whether it is indeed possible to attach a single-number valuation to an endowment as complex and as various as intelligence; second, a biologically well-founded feeling of repugnance to the notion that differences of intelligence are to so high a degree under genetic control that all the apparatus of pedagogy and special training is necessarily relegated to an altogether minor role. To these have recently been added a third, some grave doubts about the probity of Cyril Burt's investigations of intelligence quotients in twins—researches which led him to conclusions which have had a profound and by no means wholly beneficent effect on educational theory and practice. Burt's work has been the subject of extensive correspondence and annotation in both the London *Times* and the *Sunday Times*.

We must consider first the illusion embodied in the ambition to attach a single-number valuation to complex quantities—a problem that has vexed demographers in the past, and also soil physicists—as Dr J. R. Philip, FRS, has pointed out.[1] It bothers economists, too.

Although the more disputations IQ psychologists give the impression of being incapable of learning anything from anybody, it seems only fair to give them a chance not to persist in the errors of judgement that have been avoided in so many other areas of learning. Let us discuss the single-number valuation of complex variables in a number of different contexts.

First, demography. In the days when it was believed that the people of the Western world were dying out through infertility, it was thought an obligation upon demographers to devise a single-value measure of a nation's reproductive prowess and future population prospects. Kuczynski accordingly offered up his 'net reproduction rate' and R. A. Fisher and A. J. Lotka the 'Malthusian parameter' or 'true rate of natural increase'. Both had their adherents, and confident predictions were based on both, but the predictions were mistaken and today no serious demographer believes that a single-number valuation of reproductive vitality is feasible: reproductive vitality depends on altogether too many variables, not all of which are 'scalar' in character. Among them are the proportions of married and of unmarried mothers, the prevailing fashions relating to marriage ages, family numbers and the pattern of family building, the prevailing economic and fiscal incentives or disincentives to procreation, and the availability and social acceptability of methods of birth control. It is no

wonder that the single-number valuations of reproductive vitality
have fallen out of use. Modern demographers now go about their
population projections in a biologically much more realistic way,
basing them essentially upon the sizes of completed families and
the analysis of 'cohorts'—groups of people born or married in one
specific year.

Somewhat similar considerations apply to the attempt to epit-
omize in a single figure the field behaviour of a soil. The physical
properties and field behaviour of soil depend upon particle size
and shape, porosity, hydrogen iron concentration, bacterial flora,
and water content and hygroscopy. No single figure can embody
in itself a constellation of values of all these variables in any single
real instance.

Rather similar considerations apply to the way some economists
use the notion of GNP ('the tribal God of the Western world').
GNP as such may be an unexceptionable idea, but there has been
an increasing tendency to use the growth rate of GNP, positive or
negative, as a measure of national welfare, well-being, and almost
of moral stature. Any such use is, of course, totally inadmissible:
how can a single figure embody in itself a valuation of a nation's
confidence in itself, its practical concern for the welfare of its
citizens, the stability of its institutions, the safety of its streets, and
other such non-scalar and therefore presumably unscientific
variables?

IQ psychologists would nevertheless like us to believe that such
considerations as these do not apply to them; they like to think
that intelligence can be measured as if it were indeed a simple
scalar quantity. I recall in particular the barefaced impudence
with which a notorious IQ psychologist has proposed that a person's IQ *is* his intelligence as much as his height might be 5 feet
5 inches. Unhappily for IQ psychologists, this is not so. If they
were merely playing an academic game that did not affect the rest
of us for good or ill, they would of course be entitled to define
intelligence in any way they wished, but for the educated,
'strength of understanding', as Jane Austen described it, is a
complicated and many-sided business. Among its elements are
speed and span of *grasp*, the ability to see implications and con-

versely to discern a *non sequitur* and other fallacies, to discern analogies and formal parallels between outwardly dissimilar phenomena or thought structures, and much else besides. One number will not do for all these, even if—to take what must surely be one of the most abject of arguments put forward by IQ psychologists in favour of single-value mensuration—a child's IQ score is positively correlated with his income in later years.

To turn now to the vexed problem of the heritability of intellectual differences, it may be said with some confidence that unless intellectual abilities are unlike all others and unless human beings are unlike all other animals in respect to possessing them—two suppositions that are by no means as far-fetched as we may at first incline to think them (see below)—then intellectual differences are indeed genetically influenced. This applies even if upbringing and indoctrination are of preponderant importance: for here we should certainly expect inherited differences in teachability and the ability to profit by experience.

The subject is bedevilled more than any other by the tendency of disputants to spring into political postures which allow them no freedom of movement. Thus it is a canon of high Tory philosophy that a man's *breeding*—his genetic make-up—determines absolutely his abilities, his destiny and his deserts; and it is no less characteristic of Marxism that, men being born equal, a man is what his environment and his upbringing make of him. The former belief lies at the root of racism, Fascism, and all other attempts to 'make Nature herself an accomplice in the crime of political inequality' (Condorcet), and the latter founders on the fallacy of human genetic equality ('A strange belief', said J. B. S. Haldane—though a long-time member of the CP).

Confronted with this dilemma, modern liberals are keenly aware that, not so very long ago, there were countries in which those who questioned the dogma of genetic élitism would have been trampled down by big boots; but they have been slow—as liberals sometimes are—to realize that today it is the other way about; those whose views conflict with the dogma of equality are vilified, shouted down and rebutted by calumnies. Human geneticists are particularly vulnerable to the vilification of doctrinaire

Marxists because, as scientists, they are in thrall to such bourgeois superstitions as the desirability of telling the truth. Among the latest victims of such vilification are the human geneticists engaged in human karyotype screening, which entails the investigation of the human chromosome make-up at birth or earlier, to identify in good time such abnormalities as are now known to be associated with Down's syndrome ('mongolism'), a number of disorders of sexual development (for example, Turner's syndrome, Klinefelter's syndrome), and sometimes grave personality disorders, particularly that which is associated with the human sex chromosome make-up symbolized as 47XYY. The president of the American Society of Human Genetics, Dr John L. Hamerton, delivered a wise and temperate address on the problems raised by karyotype screening at the annual meeting of the society in Baltimore in 1975.[2]

Chromosomal abnormalities are unfortunately irremediable, but this is not to say that, with advance warning, their physical and behavioural consequences cannot be the subject of meliorative or preventive intervention. Nevertheless, malevolent intentions are taken for granted by disputants claiming to speak—as they all do—for 'the people'.

The really important question, however, is whether or not it is possible to attach exact percentage figures to the contributions of nature and nurture (Shakespeare's terminology) to differences of intellectual capacity. In my opinion it is *not* possible to do so, for reasons that seem to be beyond the comprehension of IQ psychologists, though they were made clear enough by J. B. S. Haldane and Lancelot Hogben on more than one occasion, and have been made clear since by a number of the world's foremost geneticists.

The reason, which *is*, admittedly, a difficult one to grasp, is that the contribution of nature is a function of nurture and of nurture a function of nature, the one varying in dependence on the other, so that a statement that might be true in one context of environment and upbringing would not necessarily be true in another. To choose an extreme example: the low-grade mental deficiency known to be associated with a constitutional inability to handle the dietary ingredient phenylalanine is a departure from normality that might be judged simply hereditary in children brought up on a normal diet abundant in the dietary constituent they cannot handle; for phenylketonuria is certainly due to a conjunction of genes that are inherited according to straightforward Mendelian rules.

If, however, a newborn child with the make-up that would otherwise have made it a victim of phenylketonuria is brought up in a microcosm free from phenylalanine—a difficult and expensive feat—then phenylketonuria would not make itself apparent. In this extreme case therefore a situation can be envisaged in which the disability is wholly environmental in origin. It will manifest itself in the presence of phenylalanine but not in its absence, and will thus present itself as a disease caused by phenylalanine. Alternatively, in a real world abundant in phenylalanine we can confidently describe the departure from normality as genetic in origin.

This example is perhaps too extreme to be informative, so I shall use instead an example which may help to make the point more clearly. The little brackish water shrimp *Gammarus chevreuxi* is extruded from the brood pouch with red eyes, but usually ends up with black eyes—because of the deposition in them of the black colouring matter melanin. The capacity for forming melanin and the rate at which it is formed and deposited are between them under the control of a number of genetic factors. Colouration of the eye is also affected by a number of other environmental factors: certainly the temperature and probably (though I don't know for sure) the dietary availability of such substances as tyrosine and phenylalanine or their precursors.

Among these various factors temperature is perhaps the most instructive, for it is possible to choose a genetic make-up such that colouration of the eye will appear to be wholly under environmental control: black at relatively high temperatures of development and reddish or dusky at lower temperatures. It is also possible to choose an ambient temperature at which red eyes or black eyes are inherited as straightforward alternatives according to Mendel's laws of heredity. Thus to make any pronouncement about the determination of eye colour it is necessary to specify both the genetic make-up and the conditions of upbringing: neither alone will do, for the effect of one is a function of the effect of the other. It would therefore make no kind of sense to ask what percentage of the colouration of the eye was due to heredity and what percentage was due to environment.

In an earlier paragraph I referred to the extreme likelihood of heredity's playing some part in the determination of differences of intellectual performance, adding, however, for form's sake, the qualification 'unless intellectual abilities are unlike all others and unless human beings are unlike all other animals in respect to possessing them'. This possibility I should now like to consider in the light of modern ethological research and our newer philosophic understanding of the character of cultural inheritance in mankind.

Human beings owe their biological supremacy to the possession of a form of inheritance quite unlike that of other animals: exogenetic or exosomatic heredity. In this form of heredity information is transmitted from one generation to the next through non-genetic channels—by word of mouth, by example, and by other forms of indoctrination; in general, by the entire apparatus of culture. I have illustrated this idea by pointing out that it was not the making of a wheel that represented a characteristically human activity, but rather the communication from one person to another and therefore from one generation to the next of the know-how to make a wheel. In this view, Man is not so much a tool-making as a communicating animal. Exogenetic or cultural heredity is that which has made possible the inauguration and retention of the cultural and institutional elements of our current civilization.

Apart from being mediated through non-genetic channels, cultural inheritance is categorically distinguished from biological inheritance by being Lamarckian in character; that is to say, by the fact that what is learned in one generation may become part of the inheritance of the next. This differentiates our characteristically human heredity absolutely from ordinary biological heredity, in which no specific instruction can be imprinted upon the genome in such a way as to become part of the package of inheritance: in ordinary evolution genetic processes are *selective* and not *instructive* in character: genetic changes do not arise in response to an organism's needs and do not, except by accident, gratify them. There is no great mystery about what has made this new pattern of heredity and evolution possible: it has been made possible by the evolution of an organ, the brain, of which the main function is to receive information from the environment and to propagate it. In such a system of heredity, indoctrination on the one hand and on the other hand imitation ('aping') and teachability play crucially important parts—as they are already known to do in the behaviour of cats and of apes.

It is very likely therefore that selective forces acting on mankind will have promoted the power of the brain to receive and communicate information, and will have made teachability an endowment of premier importance, so that, while there are likely to be inherited differences of teachability, it is extremely unlikely that teaching and training cannot improve intellectual performances. Indeed, if an intellectual performance were to be totally unaffected by training and practice I should be inclined to think that the wrong performance was being measured. It is because of the embarrassingly foolish belief that an IQ performance measures a person's 'innate intelligence' that extreme hereditarians take the view that IQ is invariant under educative procedures—a claim that reminds one of Francis Galton's contempt for those who try to raise themselves beyond whatever station in life it may have pleased their genes to call them to. If it were indeed true that IQ is invariant with age then the only conclusion we could legitimately come to is that the tests upon which its measurement is based are tests of the wrong capacities.

In short, although the possibility of its being so was introduced more as a formal disclaimer than with any other serious purpose, we can conclude that the pattern of inheritance of intellectual differences in human beings is indeed different from the inheritance of other character differences in other animals.

In his *The Science and Politics of IQ*,[3] Leon J. Kamin is 'concerned with a single major question: are scores on intelligence tests (IQ's) hereditable?'. The answer, he says,

in the consensus view of most intelligence testers, is that about eighty per cent of individual variation in IQ scores is genetically determined. This is not a new conclusion. Pearson, writing in 1906, before the widespread use of the IQ test, observed that 'the influence of environment is nowhere more than one-fifth that of heredity, and quite possibly not one-tenth of it.' Herrnstein, reviewing the history of intelligence testing to 1971, concluded 'We may therefore say that 80 to 85 per cent of the variation in IQ amongst whites is due to the genes.'

Kamin goes on to state it as a principal conclusion of his book that: 'There exist no data which should lead a prudent man to accept the hypothesis that IQ test scores are in any degree heritable', and then asks how it is that a contrary opinion has so long prevailed among psychologists. Kamin himself believes that sociopolitical motives underlie the willing assent of IQ psychologists to the notion of inherited differences in intellectual capacities. Indeed, he carries this conspiracy theory of heritability to the point of suggesting that the entire project of IQ psychology is implicitly a great salve for the public conscience and incidentally a great relief to the public purse: if the poor are unsuccessful and inferior because they have been born that way and not because of the way they have been treated, then there is not much we can do about it.

Thus the extreme hereditarian viewpoint is seen as part of that great conspiracy referred to above to 'make Nature herself an accomplice in the crime of political inequality'. The conspiracy is not, of course, declared and open, but is rather the subconscious consequence of these economic and class-competitive forces that are thought to shape history. Thus Kamin's interpretation of the origins of hereditarian theory has about it the kind of Olympian glibness more often found in psychoanalytic theory, and it is equally difficult to refute. For just as any criticism of psychoanalysis is construed as an infirmity of the psyche which itself requires psychoanalytic treatment, so criticism of an essentially Marxist theory is thought to reveal its author as yet another victim and dupe of the very socio-economic forces whose efficacy he has presumed to question.

In writing of the pioneers of IQ testing, Kamin makes the useful point, quite new to me, that when Alfred Binet pioneered intelligence testing he described as 'brutal pessimism' the belief that the intellectual performance of an individual could not be augmented by special training, and indeed prescribed a therapeutic course in 'mental orthopaedics' for those with lowly test scores.

Binet was an agent of the State schooling system in France, and the purpose of his intelligence tests was to identify children in need of special schooling. It was far otherwise with Binet's American heirs, particularly Lewis M. Terman, who came to regard an intelligence test score as a measure of a fixed quality thought of as 'innate intelligence'—an expression still in use and as clearly indicative today as it ever was of a deep-seated misunderstanding of genetics. Moreover, a political, racist and—in the worst sense of the word—eugenic motivation is made painfully clear by some of Kamin's quotations from the pioneers of IQ testing. They may not have been the worst offenders: they were writing at a time when it was widely believed that the riotous proliferation of the feeble-minded would repopulate the world with imbeciles, and that affections such as 'mongolism' (Down's syndrome) represented an atavistic degeneration to a primitive and lowly human type (hence the name).

Nevertheless, the alleged malevolence, racial bias or even downright dishonesty (see below) of hereditarian psychologists cannot answer the material question whether or not heredity contributes anything to differences of intellectual performance. In denying any such influence, Kamin goes too far—just as H. J. Eysenck went too far in a passage the mere contemplation of which probably now causes him acute embarrassment: 'the whole course of development of a child's intellectual capabilities is largely laid down genetically'.

With thinkers such as Terman to guide them, we need not wonder at how nearly castration became a statutory requirement in a number of American States, nor at how confidently one State legislature or another resolved that heredity played a principal part in crime, idiocy, imbecility, epilepsy and dependence on charity.

The first tests to reveal that Blacks score less than Whites emerged from the extensive screening undertaken in the First World War, tests of which Kamin drily remarks that they 'appear to have had little practical effect on the outcome of the war'. Such tests have, however, had a profound effect on the relationship between Blacks and Whites ever since. Another important part of the harvest of the routine screening of recruits was a vast heap of unreliable information on the intelligences of recruits classified by their countries of origin—evidence from which it became pretty clear that northern European countries scored highest, with Mediterraneans, Slavs and other such lowly types a good way behind. These findings became known in Congress and had an important influence in shaping the US immigration laws.

Madison Grant, in his *The Passing of the Great Race*, lamented the likelihood that the American people would be irretrievably diminished by the influx of inferior foreigners. With his nice touch for allowing the subjects of his criticism to assassinate themselves, Kamin quotes passages from Grant and a Professor C. C. Brigham of Princeton that sound like a crash course in racism:

The Nordics are . . . rulers, organizers, and aristocrats . . . individualistic, self-reliant, and jealous of their personal freedom . . . as a result they are usually Protestants . . . The Alpine race is always and everywhere a race of peasants . . . The Alpine is the perfect slave, the ideal serf . . . the unstable temperament and the lack of coordinating and reasoning power so often found among the Irish . . . We have no separate intelligence distributions for the Jews . . . Our army sample of immigrants from Russia is at least one half Jewish . . . Our figures, then, would rather tend to disprove the popular belief that the Jew is intelligent . . . he has the head form, stature, and color of his Slavic neighbors. He is an Alpine Slav.

It seems to me that many of the socially disruptive influences that have been drawn from the study of IQ performances are the consequence not so much of the malevolence of those who undertake them as of the inherent failings of IQ testing itself. The illusion that a single-number valuation can be attached to anything that an educated man means by the word 'intelligence' has already been exposed. But a still graver illusion is even more dangerous because it places foreigners, the poor and the deprived at a special disadvantage—the illusion that intelligence tests can be devised which are 'culture free', that is, which are quite uninfluenced by the cultural background of the subject's family, or by the linguistic or performative exercises which he may, or more likely may not, have taken before testing. These naïve beliefs are now passing out of favour, but not before they have done a very great deal of harm.

Where so many hereditarian writers are graceless, rancorous and inept, Kamin writes with a winning skill that Jonathan Swift would have delighted in. He is merciless to the Californian sickness:

> The meek might inherit the kingdom of Heaven, but, if the views of the mental testers predominated, the orphans and tramps and paupers were to inherit no part of California. The California law of 1918 provided that compulsory sterilizations must be approved by a board including 'a clinical psychologist holding a degree of Ph.D.'. This was eloquent testimony to Professor Terman's influence in his home state.

These passages of fine polemical writing must not be allowed to distract attention from the most important part of Kamin's book, his critique of observations purporting to demonstrate a very high degree of heritability of differences in IQ scores. Because it has been the subject of some searching investigative journalism by the *Sunday Times* in London, we shall pay particular attention to the testing of twins.

Kamin gives an admirably lucid account of the methodology of twin studies, of which the underlying principle is this: twins may be of the kind called identical, that is, the product of a single fertilized egg, or they may be 'fraternal', that is, litter mates—who resemble each other genetically no more closely than ordinary brothers and sisters. Identical twins can be assumed fairly confidently to have the same genetic make-up. Identical twins who have been separated and brought up in different environments are methodologically a godsend. The degree of correlation between their measured intellectual performances is an estimate of the degree to which heredity has contributed to them, *provided* the various environments are representative of the whole range of environments to which human beings are exposed, and twins themselves are representative of the entire population of which they are members.

However, as Kamin writes, 'there is little reason to suppose that these assumptions hold in any of the studies that have been made of separated twins'. Kamin pays special attention to the studies made by Professor Cyril Burt, one of the great pioneers of educational psychology, and Professor of Psychology at one of England's three leading universities: University College, London. Burt's direct influence was probably, largely, a harmful one; because of his advocacy and the tendency to regard his opinions as Holy Writ, eleven-year-olds in Great Britain were subjected to tests intended to divide the bright from the comparatively dull. Indirectly his teachings may be said to have invited the backlash which has led now to the reinstitution of those comprehensive schools that are founded on the proposition that all children are fundamentally of equal ability—so making the usual confusion between the fact of biological inequality and the political right to equal treatment. Still, he can hardly be blamed if, for political reasons, his teachings have now had the effect of handicapping those very children whose interests they were designed to promote.

Kamin's criticisms of Burt make some of the most damaging accusations that can ever have been levelled against a scholar:

the various papers published by Burt often contain mutually contradictory data, purportedly derived from the same study. These contradictions, however, are more than compensated for by some remarkable consistencies which occur repeatedly in his published works. The first examples that we shall cite do not involve his study of separated twins, but later examples will do so.

The papers of Professor Burt, it must be reported, are often remarkably lacking in precise descriptions of the procedures and methods that he employed in his IQ testing. The first major summary of his kinship studies, a 1943 paper, presents a large number of IQ correlations, but virtually nothing is said of when or to whom tests were administered, or of what tests were employed. The reader is told, 'Some of the inquiries have been published in LCC reports or elsewhere; but the majority remain buried in typed memoranda or degree theses.'

Toward the end of 1976 a furore was started by the publication in the London *Sunday Times* of an article by a team of investigative journalists led by Dr Oliver J. Gillie, their medical correspondent and a gifted geneticist. The investigations questioned the probity of Burt's entire work, raising a number of awkward questions to which no satisfactory answers had then been given. In addition, Professor Jack Tizard, the highly respected Professor of Child Development in London University, delivered a lecture likening the revelations about Burt to those which disclosed the forgery of the Piltdown skull. Tizard said his suspicions had been aroused two and a half years beforehand by his complete failure to find two people at University College who were said to have worked very closely with Burt in his research—colleagues with whom he had, indeed, published a number of papers between 1952 and 1959—namely Miss Howard and Miss Conway.

The *Sunday Times* team fared no better; they could find no record that either had ever been on the staff of the Psychology Department at the University College, and could not even trace them in the files at Senate House, reputedly the central nervous system of the University of London, which holds duplicates of the documents of the University's constitituent colleges: 'Direct inquiries to 18 people who knew Burt and his circle well from the 1920s, when he was at the National Institute of Industrial Psychology, until he died, have failed to find anyone who met Howard or Conway or knew of them, and no one with these names is listed in the files of the British Psychological Society.' The *Sunday Times* concluded its investigation by proposing the hypothesis that Misses Howard and Conway never existed.

However, in spite of these misgivings there seems no doubt that Margaret Howard anyway did 'in some real sense' (as philosophers say) exist. Professor John Cohen of Manchester University is quoted by Dr Gillie as saying that he knew Miss Howard well. In the follow-up article containing this revelation he also quotes a damaging accusation by Professor and Mrs Clarke of Hull University that articles which they did not write were published in their

name by Sir Cyril Burt, and they add, 'It is extremely difficult to
see how Burt arrived at some of his figures on inheritance of
intelligence without cooking them.'

Kamin's evidence and the *Sunday Times* investigations point to
Burt's having a fairly lofty attitude toward the provenance
and probity of his empirical data. Indeed, the accusation that
Burt's findings were too good to be true—that is, were too closely
in line with theoretical anticipations—gives us a clue to the
most puzzling question of all about Burt: Why, why did he act
deviously?

The only explanation I can think of is that a belief in the
predominant influence of heredity in relation to intellectual performances has the same kind of appeal for those who hold it as
Lamarckism—the belief in an inherence of acquired characters—
has had for unskilled biologists. It seems to them so *right*, so
obviously and necessarily true, so clearly in keeping with their
sense of the fitness of things, that people who do not share their
beliefs must somehow be persuaded in their own best interests to
do so, if necessary by a slight adjustment of the figures here, an
assumption based upon a lifetime of experience there, and judicious selection of data somewhere else. Fraudulent experiments
have been used to uphold Larmarckian interpretations of heredity, and in Burt's methodological malpractices we may have another case in point. Villainy is not explanation enough: Burt
probably thought of himself as the evangelist of a Great New
Truth.

So much anyway for the case for the prosecution. The most
significant utterance in the case for the defence is that of Professor
Eysenck, himself a dedicated hereditarian. In a letter in the *Sunday
Times* citing Burt's data and calculations he concedes that some of
his procedures were 'of course inadmissible' to a degree 'that
makes it impossible to rely on these figures in the future'.[4] On the
other hand Professor A. R. Jensen, joining in the *Times* debate, did
not share Eysenck's view that any of Burt's procedures were
inadmissible: he dismisses the attack on Burt as so much calumny
and concludes, with the 'complete confidence' which natural
scientists so seldom feel, that 'even if all of Burt's findings were

thrown out entirely, the picture regarding the heritability of IQ would not be materially changed'. I am quite sure Jensen is not intending to be ironical; but this judgement does seem to be a rather strange compliment to a man thought of as a founding father of psychometry.

There is, as a matter of fact, a well-established precedent for the selection or adjustment of figures to fit a preconceived hypothesis: R. A. Fisher, at that time the world's foremost authority on small-sample statistics, once pointed out that Mendel's famous segregation ratios (3:1; 9:3:3:1) were numerically much too good to be true. Given the size of his samples, no such degree of conformity to theoretical anticipation could be judged plausible. Whatever R. A. Fisher's motives may have been in calling attention to this fact, we may be quite sure it was not his intention to show Mendel up as a running-dog of Fascism (as the faithful later came to call him). The most plausible explanation seems to be that the abbé's gardeners and assistants had formed a pretty clear idea of what ratio Mendel was expecting, and whether out of loyalty or affection supplied their reverend employer with results they thought he would like to hear.

There is, however, a profoundly important difference between the cases of Mendel and of Burt: Mendel was right.

Now that the IQ controversy has risen to a new height and shows no sign of abating, the publication of Block's and Dworkin's *The IQ Controversy*[5] is particularly timely and valuable. It is in the genre known as a 'reader', that is to say it gives us a conspectus of prevailing opinions in the words of those who hold them. The danger of a reader in such a context as this is that the editor may, by judicious selection or omission, prejudice the conclusions that an impartial reader might come to. Block and Dworkin have not done this: their editorial matter provides only that minimum of connective tissue which a book such as this urgently needs. It is also fair to point out that the only way in which the hereditarians could be rescued completely from public obloquy would be by omitting their contributions altogether. Block and Dworkin have rightly decided against so partial a procedure: they play fair, though it might be thought cruel to republish the controversy between Walter Lippmann and Lewis Terman, published in the *New Republic* in 1922 and 1923. This gives one the sick feeling that people of sensibility have when they first witness a bullfight: the contest is so cruelly unequal when one contestant has nothing but a slow-footed and ponderous irony with which to defend himself against the highly intelligent, light-footed and cruelly provocative Mr Lippmann.

A special strength of their book—and one that enormously enhances its value for college reading—is their generous allocation of space to such real professionals as Richard Lewontin and John Thoday, with a passing quotation from Michael Lerner. In the course of a grave, learned and witty investigation of what has come to be called 'Jensenism', Lewontin remarks: 'There is no such thing as *the* heritability of IQ, since heritability of a trait is different in different populations at different times. Second, the data on which the estimate of 80 per cent for Caucasian populations is based, are themselves of very doubtful status.'

The citation from Michael Lerner includes this sentence: 'it is a fact that generations of discrimination have made direct comparisons of mental traits between Negroes and whites not biologically meaningful'.

John Thoday expounds clearly and critically the methodology of intra-group and between-group comparisons, calling attention, as he does so, to blunders by IQ psychologists of a kind that disclose a truly deep-seated misunderstanding of genetic principles. He concludes that 'there is no evidence which reveals whether the Negro–white IQ difference has any genetic component or any environmental component.'

The reflection that might well be in the forefront of the minds of laymen as they put down the Block and Dworkin book is this: The question of the heritability of differences of IQ is one upon which everybody feels entitled to have an opinion. In recent years even a prominent electrician has felt authoritative enough to have his say; yet on matters to do with heritability it might be thought prudent to give most weight to the opinions of geneticists. Why, then, is it that some of the world's most prominent geneticists— among them Michael Lerner, Richard Lewontin, Walter Bodmer and John Thoday—remain so deeply unconvinced by the hereditarian arguments of such as Jensen and Eysenck? We need not resort to murky ideological explanations to find the reason. It is more likely, I suggest, that at a time of deeply troubled race relations, when the whole possibility of peaceful co-existence and mutual respect in multiracial communities is in question in many parts of the world, these geneticists feel an imperatively urgent desire to put the scientific record straight.

14 Florey story

Howard Walter Florey was a great man and no mistake. He devoted the more important part of his professional life to a single wholly admirable purpose which he pursued until he achieved it, showing, in spite of many setbacks and rebuffs, the magnanimity that is the minimal entry qualification for being considered 'great'. In a memorial address, Patrick Blackett likened Florey's achievement to that of Jenner, Pasteur, and Lister: but the public were so little aware of him that when Macfarlane first approached publishers with the notion of a biography,[1] they wondered if he would not do better to write on Alexander Fleming instead. This, Macfarlane surmises, was because the public had already cast Fleming as the hero of the great penicillin story: he was a closer approximation than Howard Florey to the public's stereotype of a great scientist, for, although a great scientist, Florey was the kind of man who would have been a success at anything he had chosen to turn his hand to. Macfarlane thinks the comparison of Florey with Jenner, Pasteur, and Lister is specially apt because 'the work of these three men forms a logical sequence with his own that spans, in the course of about 150 years, the gulf between almost total therapeutic helplessness and the virtual defeat of most of the important bacterial diseases.' Whatever the general public may have thought about him, Florey stood unsurpassably high in the estimation of his colleagues—that which means most to a scientist—and in due course they elected him head of their profession in England as President of the Royal Society.

It is fortunate for mankind that no Geneva convention prohibits the prosecution of germ warfare by germs themselves, among whom the struggle for existence is murderous and unremitting. Penicillin is one of a class of substances manufactured by moulds and bacteria—particularly soil bacteria, which live in deplorable conditions of squalor and overcrowding: these are substances which suppress the growth or multiplication of other microorganisms. Penicillin was discovered by Alexander Fleming: by luck, so it is believed, though in reality Fleming had been looking for something very like penicillin all his life. The only element of blind luck about the discovery of penicillin was that, unlike most antibiotics, penicillin is not poisonous to human beings and other higher animals: the reason is that penicillin interferes with metabolic processes peculiar to bacteria, whereas some other antibiotics like actinomycin and mitomycin are toxic because they obstruct a cellular activity common to bacteria and ourselves.

Before a good scientist tries to persuade others that he is on to something good, he must first convince himself. The first experiment that convinced Florey and his two colleagues, Norman Heatley and Ernst Chain, that they might be on to something occurred very shortly after the German Army—'to the inexplicable surprise of the Allied Command—instead of dashing themselves to pieces on the Maginot line, drove their Panzer columns round the northern end of it, and swept between the British and French armies against almost no resistance until they reached the coast near Abbeville'. At 11 a.m. on Saturday, 25 May 1940, eight white mice received approximately eight times the minimal lethal number of streptococci. Four of these were set aside as controls, but four others received injections of penicillin—either a single injection of 10 milligrams or repeated injections of 5 milligrams.

The mice were watched all night (but of course). All four mice unprotected by penicillin had died by 3.30 a.m. Heatley recorded the details and cycled home in the black-out. Poor mice? Yes of *course* poor mice, but poor human beings too, don't forget:

Next morning, Sunday 26th May, Florey came into the department to discover that the results of his experiment were clear-cut indeed. All four control mice were dead. Three of the treated mice were perfectly well; the fourth was not so well—though it survived for another two days. Chain arrived, and then Heatley, who had had very little sleep. They all recognised that this was a momentous occasion. What they said is not recorded, but memory has supplied subsequent writers with various versions. One might suppose that Heatley said very little, that Chain was excited, and that Florey's reported comment 'It looks quite promising' would be entirely in character.

Animal experiments on a much larger scale soon made it clear that penicillin was indeed of great potential importance. The first published paper on the subject in the *Lancet*, by Chain, Florey, Gardner, Heatley, Jennings, Orr-Ewing, and Sanders—the names are in alphabetical order—stated: 'The results are clear-cut and show that penicillin is active *in vivo* against at least three of the organisms inhibited *in vitro*.'

Macfarlane's account of the animal experiments and the first clinical trial is simple and straightforward, and all the more exciting for being so. It makes my heart pound still although I know the outcome, for the thrill of reading about these great occasions does not diminish: scientists are like cricket-lovers who never tire of reading or recounting what Colin Milburn used to do to short balls on the leg side from Wesley Hall.

Florey's and his colleagues' clear awareness of its importance raised the problem of what would become of their strain of *Penicillium* if—as seemed entirely on the cards—the German Army were to reach Oxford and sack it: 'The mould itself must be preserved, undetected. Florey, Heatley, and one or two others smeared the spores of their strain of *Penicillium notatum* into the linings of their ordinary clothes where it would remain dormant but alive for years.'

Because of its potency and non-toxicity penicillin is the paradigm of antibacterial substances, but it is not without snags: the warfare between germs which, as I suggested above, leads to the formation of substances like penicillin leads also to the evolution of remarkably effective mechanisms of defence. One is the manufacture of the ferment *penicillinase* which destroys penicillin and thus protects bacteria from it. The widespread use of penicillin—sometimes injudiciously often—has led to the evolution in many hospitals of strains of bacteria resistant to its action: once a mutant impervious to penicillin has arisen, natural selection soon brings it about that the mutant becomes the prevailing type in the population. It is not that penicillin has lost any virtue, but rather that bacteria have acquired a vice.

Another snag, exacerbated by the tendency of clinicians in the early days of penicillin to administer colossal intramuscular doses in the presence of substances known to immunologists as 'adjuvants', is that penicillin can give rise to severe allergic reactions in a specially susceptible minority of those who receive it. The development of new antibiotics or new variants of penicillin has gone a long way to annul this disadvantage.

Among the difficulties associated with the production and use of penicillin, I have mentioned only the evolution of penicillin-resistant strains of bacteria, and the ability of penicillin to sensitize susceptible patients. The greatest snag of all was the sheer difficulty of producing quantities sufficient for clinical trials. But Florey's School of Pathology became a pilot-scale production plant for the purpose, and it was natural that he should turn to the great pharmaceutical companies to make use of their great practical experience and know-how. One of the American companies, Merck, knew how only too well. According to Macfarlane's account, Norman Heatley's visit to share with them what he knew about the production of penicillin was marked by something much less than candour on the part of Merck, who had prepared applications for British and American patents covering the essential stages of production processes devised by the Oxford scientists and one of their own scientific officers: 'This fact was not generally appreciated until 1945, when British firms discovered that they had to pay royalties on their penicillin production.' Whatever their moral shortcomings, the great pharmaceutical companies did in due course produce penicillin in adequate amounts—the consideration that mattered most. The complexity of the production of penicillin and the murkiness of its origin ('see *Macbeth*, Act Four, Scene One,' said Oxford wags not unwilling to poke fun at a discovery so obviously important) impeded the funding of Florey's research, for in the 1930s Gerhard Domagk's discoveries had ushered in the era of the sulphonamides, also powerful antibacterial agents. They, being synthetic organic chemicals, could be produced without cauldrons and the toil and trouble that go with them. It is clear that some know-alls serving on the Medical Research Council at the time must have resolved that the future of antibacterial therapy lay with these synthetic organic chemicals and not with 'biologicals' such as penicillin, for, much to the annoyance of Florey and Chain, the MRC did not fund penicillin as handsomely as the occasion called for. Luckily, however, the Rockefeller Foundation helped out. The sums involved—of the order of hundreds of pounds—seem comically small by modern standards, but money went much further then.

A leading article in *The Times* on 'Penicillium' referred, without mentioning any names, to the research in progress in Oxford. Sir Almroth ('stimulate the phagocytes!') Wright addressed the Editor thus about his former pupil:

Sir,
    In the leading article on penicillin in your issue yesterday you refrained from putting the laurel wreath for this discovery round anyone's brow. I would, with your permission, supplement your article by pointing out that, on the principle *palmam qui meruit ferat*, it should be decreed to Professor Alexander Fleming of this laboratory [St Mary's Hospital]. For he is the discoverer of penicillin and was the author also of the original suggestion that this substance might prove to have important applications in medicine.

Ever since Wright's letter, there have been attempts to make a *cause* out of the allocation of credit for the great discovery: first by comparing the contributions of Fleming and Florey, and more recently by comparing the contributions of Florey and Sir Ernst Chain. But no journalist will get any copy out of Macfarlane: he treats the whole subject wisely and temperately, as might be expected of a historian who is a distinguished scientist.

    Human nature, unfortunately, is such that so great a discovery as that in which Fleming and Florey played crucially important parts is certain to be followed by jealous attempts to diminish it by finding evidence that it had all been thought of or done before.

Certainly Pasteur recognized that germs engaged in germ warfare, and maybe the Chinese *did* put mouldy soya bean curds to therapeutic uses: but Alexander Fleming discovered penicillin, and Howard Florey was the prime mover in turning it into the most important therapeutic innovation of the twentieth century. Both were necessary, but neither can be judged singly sufficient.

Relations between Florey's team and Fleming were inevitably difficult. Florey did what was proper: that is to say, he acknowledged Fleming as *the* discoverer of penicillin in his first paper on the subject, but Fleming always felt he deserved more credit than he got, and referred often to 'my brainchild'. 'What have you been doing with my old penicillin?' Fleming asked the Oxford team when he came down to visit Florey's laboratory. Florey and Chain told him what they were doing and took him on a tour of the laboratory: 'Fleming said almost nothing during this inspection and returned to London without comment or congratulation on what had been achieved.' It is clear from Ronald Hare's *The Discovery of Penicillin* (1970), upon which Macfarlane draws gratefully, that Fleming was an amateur in the big business of practical therapeutics, where Florey was every inch a pro.

Florey was the greatest experimental pathologist of his day. Penicillin was not his only—nor even his principal—interest: to judge by the quietly passionate persistence with which he studied the problem and persuaded all his young colleagues to do so too, the central interest of his scientific maturity was to elucidate what came to be called 'the great lymphocyte mystery'. Lymphocytes are white blood corpuscles—those, as we now know, that transact immunological reactions. The lymphatic vessels of the body, which drain fluid from the tissues, unite into one major vessel, the thoracic duct, that empties its contents directly into the bloodstream. By this route thousands of millions of lymphocytes enter the bloodstream daily, but what becomes of them, and what is their function anyway? Whatever their intentions may have been, most of Florey's young students and co-workers—among them a future Dean of the Harvard Medical School, a future head of the Medical Research Council, and the future wife of the reviewer—found themselves trying to answer these questions. The problem was solved by Dr J. L. Gowans, the one who became head of the MRC: he found, contrary to orthodox opinion, that lymphocytes are relatively long-lived cells which circulate and re-circulate through the blood and lymph vessels. Like the chorus in a provincial production of *Faust*, lymphocytes in the bloodstream at any one moment disappear behind the scenes and re-enter by another route. Lymphocytes, moreover, are cells that manufacture antibodies and are responsible for the recognition, and ultimately the elimination, of foreign-tissue transplants and cells infected by viruses or transformed by the action of cancer-producing agents.

Another of Florey's great interests was the nature and cause of atheromatosis, the formation in blood vessels of atheromata, the waxy plaques that sometimes threaten the free passage of blood in such important vessels as the coronary arteries, which supply blood to the great muscles of the heart.

As a man of action, determined to get results, Florey might easily have made some bitter enemies, but in reality he inspired a good deal of affection and admiration among colleagues, who to this day like exchanging Florey stories and laughing at their discomfiture when some characteristically sardonic or scathing remark of his cut them down to size. I wrote a long-winded paper on my work as a graduate student in his lab and showed it to Florey. When he handed it back to me, he said: 'I don't see what you're getting at, Medawar. The paper doesn't make sense to me.' Later on, having learned better, I wrote a clear and simple paper for a journal Florey was particularly fond of and regularly read, and I was overjoyed when Florey passed me in one of the narrow lanes that wind through the science area in Oxford, twitted me in his usual style on having rushed precipitately into print, but added: 'Your paper's not at all bad, Medawar.'

Florey was extremely intelligent, clear-sighted and shrewd, but he was not an intellectual, and even at the height of his success he was a tiny bit scared of such people, for he was not always as sure of himself as might have been expected of a man so enormously successful. 'Was that all right?' Florey once said to me after giving an important lecture at the Royal Society, of which he was President. It was, but I thought it touching and endearing that he should still want to be assured of the good opinion of his juniors.

So deep was the impression made by Florey on his juniors that I do not believe any one of them could have written a life of Florey of which Florey would himself have disapproved. He would have liked Macfarlane's life because it is simple and straightforward and sticks to the point without clever philosophical or psychological digressions. Florey might have made some amusedly self-depreciatory remark about the use of the word 'great', but if he had done so he would have been—as he seldom was—wrong.

15 In defence of doctors

The angle of vision from a Chair of Social Medicine such as Thomas McKeown occupied with distinction for many years in the University of Birmingham, England, is quite different from that of a physician at the bedside or a surgeon at the operating table. The difference is embodied in the following credo:

I believe that for most diseases, prevention by control of their origins is cheaper, more humane, and more effective than intervention by treatment after they occur.[1]

This belief, McKeown goes on to say, 'does not reduce the importance of the pastoral or samaritan role of the doctor. In some ways it increases it.' McKeown firmly repudiates the notion that his message is cognate with that which is embodied in the 'Medical Nemesis' by the author referred to in public by the late Professor Henry Miller as 'Ivan the Terrible'.[2]

Unfortunately the antithesis between prevention and remedy as McKeown outlines it is very seldom as simple as it might at first sight appear to be as the following examples will show.

We all know very well that the frequency of the congenital affliction known as Down's Syndrome (formerly 'Mongolism' because of Down's racist propensities) would be greatly reduced if the mean age of motherhood were also to be reduced. But some women want to have—and may for one reason or another only be able to have—a child at the age of thirty or later. Again, the work of Brian MacMahon at the Harvard School of Public Health has shown very clearly that a woman who has had her first child as a teenager stands much less risk of becoming a victim of breast cancer than a woman who has had her first child in her late twenties or *a fortiori* her thirties. This finding seems to open the door to a number of salutary preventive procedures, but in real life who is going to encourage teenagers—among them one's own daughters, perhaps—to become pregnant as soon after menarche as possible to give them extra protection in later life against a misfortune that may not befall them anyway? Prophylaxis is not enough: some women will get breast cancer no matter when their children are born, just as some people who don't smoke will get lung cancer. So no matter how energetic our preventive measures, we must still have the resources of treatment at our command.

In spite of his seniority and distinction McKeown is not above being an *enfant terrible*. The philosophic doubts which form the subject of this book

began when I went to a London Hospital as a medical student after several years of graduate research in the Departments of Biochemistry at McGill and Human Anatomy at Oxford. There were two things that struck me, almost at once. One was the absence of any real interest among clinical teachers in the origin of disease, apart from its pathological and clinical manifestations; the other was that whether the prescribed treatment was of any value to the patient was hardly noticed. . . .

Living as I do in a world of medicine and medical research I am happy to be able to affirm that from my own experience what McKeown is saying is absolute bunk.

There is a good deal more in the same vein. He says that 'there seemed to be an inverse relation between the interest of a disease to the doctor and the usefulness of its treatment to the patient'. This was why, so he tells us, 'Neurology . . . attracted some of the best minds'—and that the fascination of multiple sclerosis and amyotrophic lateral sclerosis lay in diagnostic exercises that made little difference to their progress.

Since I know many who are engaged in the treatment of or research into multiple sclerosis, and since I myself do all that is in my power to promote their work, I tentatively put forward an alternative hypothesis: the interest of multiple sclerosis is that it is a terrible disease, cruelly capricious in its incidence. It arouses perhaps more than any other the feelings of compassion that play so large a part in attracting the young into the study of medicine.

It seems to me that McKeown, who temporarily casts himself in the role of St Peter, weakens his position by resolving to admit dentists into heaven. Not one of us will deny that oral hygiene and the judicious use of fluoride are much preferable to remedial dentistry. But alas, teeth decay in spite of our best endeavours, so we still need dentists—and thank God we have them.

McKeown began to think more deeply about the problems he had just become aware of when he was appointed to a Chair of Social Medicine in the University of Birmingham, his predecessor having been G. A. Auden, father of the poet. In his Chair McKeown came to see himself 'as an academic Billy Graham who bears the glad tidings of health for the taking to a grateful people'. He formed the opinion moreover that

medical science and services are misdirected, and society's investment in health is not well used, because they rest on an erroneous assumption about the basis of human health. It is assumed that the body can be regarded as a machine whose protection from disease and its effects depends primarily on internal intervention.

When McKeown finally gets down to business after an unnecessarily discursive prolegomenon he declares it as his intention to examine 'the validity of a concept . . . on which medical activities largely rest'—the concept that the maintenance of health depends upon the understanding of the structure and function of the body and the disease processes that affect it—an approach which he regards as 'mechanistic', a word which he interprets in the sense of 'machine-like' though for many years biologists have taken it to signify 'physically determinate'. There follows some uneasy discussion of the mind/body relationship during the course of which McKeown mentions G. A. Ryle and the notion of 'Category-mistakes' without giving me the impression that he altogether understands what he is talking about.

McKeown concedes that the slow 'secular' (or long-term) improvement of human health during the nineteenth century occurred *pari passu* with the growth of our knowledge about the structure and workings of the human body. But he seems impatient with the idea that the former is a consequence of the latter, for he advocates a different view: the reduction of mortality and an improvement of health in human and animal populations are due to the greater abundance and better distribution of nutriment. This transformation annulled a principal constraint upon the growth of human and other populations, a constraint dependent on population density, namely shortage of food. Serious questions can be raised about this view as we shall see.

McKeown's chapter on 'Inheritance, Environment and Disease' has a querulous and dissatisfied air throughout: teachers before the war had urged doctors to become more keenly aware than they had been until then of the gravity and prevalence of cancer of the lung, but he chides them for having paid so little attention to aetiology and to discussing the possibility 'that the disease might be due to influences which could be modified or removed'. Things are a bit better now, though, McKeown concedes: due attention is given to the importance of smoking, exercise, and diet; moreover, conscientious clinicians, by teaching and example, try to modify the practice of their students and the behaviour of their patients. 'Nevertheless in medicine as a whole the traditional mechanistic approach remains essentially unchanged; and it will remain unchanged so long as the concept of disease is based on a physico-chemical model.'

I must say I am not clear what McKeown is complaining about. If, as is possible, cancer originates as a somatic genetic accident, this is a physical event which—if it is to be understood—must be understood in physico-chemical terms. The endeavour to understand such a phenomenon is surely not incompatible with an epidemiological analysis that might help to explain its frequency. Most sensible physicians take the view that both approaches are necessary though neither is singly sufficient.

McKeown devotes a considerable number of pages to the subject of 'Inheritance, Environment and Disease'. That genetic factors control differences of susceptibility to disease is known to be true of some diseases and not known to be false of any; I look in vain through McKeown's pages to find a statement of equal clarity and there are other ways in which this chapter disappoints. It can be inferred from McKeown's discussion on the relative influences of heredity and environment that it is not in general possible to attach any one figure to the proportional contributions of the two to differences in our character makeup (e.g., in respect of IQ scores), but although McKeown allows us to draw this inference it would surely have been better if he had explained why any such exercise is impossible. It is because the contribution made by nature to a character difference is a function of nurture (and that of nurture is a function of nature).

Further, in view of McKeown's preoccupation with preventive medicine, I had reasonbly hoped for some discussion of the merits and shortcomings of the programme of J. B. S. Haldane for diminishing the frequency of 'recessive' diseases such as phenylketonuria (recessive diseases are those in which the offending gene must be inherited from *both* parents instead of—as in so-called 'dominant' diseases—from only one). The essence of the Haldane solution is the discouragement of marriage, or at all events of childbearing, by possessors of the same damaging recessive gene, a preventive measure which turns upon the fact that most victims of phenylketonuria are the offspring of a marriage between carriers of the offending gene. The shortcomings of this in many ways admirable proposal are first that carriers of recessive genes are not always identifiable and secondly that such a procedure as Haldane recommends would suspend the action of natural selection altogether and pile up still greater difficulties for future generations. Another difficulty is that putting the Haldane scheme into effect would cause a tremendous outcry from all intent upon defending the right—nay, privilege—of parents to bring into the world biochemically crippled or otherwise disadvantaged children.

The second part of McKeown's book is called 'Determinants of Health'; my spirits at once rose because McKeown is admirably well qualified to write authoritatively on the causes of the vast secular improvement in human health that has taken place over the last few centuries.

During most of man's existence it is probable that a considerable proportion of all children died or were killed within a few years of
birth . . . out of ten newborn children, on average, two to three died
before the first birthday, five to six by age six and about seven before
maturity. In technologically advanced countries today, more than 95
percent survive to adult life.

The statistical characteristic most dramatically affected by a
reduction in infantile mortality is of course the mean expectation
of life at birth. McKeown does well to point out how enormously
it has increased over the period during which reliable records of
mortality have been kept; in Sweden it rose from between thirty
and forty years in 1700 to seventy-two years for males and
seventy-seven for females in 1970. Although the available records
'leave no doubt that death rates were falling from the beginning of
the nineteenth century . . . there is impressive indirect evidence
that the decline began somewhat earlier, probably in the first half
of the eighteenth century'.

To interpret these figures McKeown says we must turn to
national records of the causes of mortality which are available for
England and Wales since 1838. In so far as it is possible to interpret the bills of mortality it seems that nearly 90 per cent of the
total reduction of the death-rate from the beginning of the eighteenth century until today can be credited to the decline of infectious disease. The different infectious diseases contributed
unequally to this decline, respiratory tuberculosis contributing
most and infections of ear, pharynx, and larynx least. The stand-
ardized death-rate from smallpox in England and Wales fell from
seventy-five per million in 1848–54 to two per million in 1971.
The corresponding figures for scarlet fever and diphtheria were
1,016 and zero. Needless to say mortality statistics do not assess
the gravity and social or personal burden of a disease; although
mortality from measles is way down, it was at one time feared—
for reasons subsequent research has not upheld—that multiple
sclerosis, surely one of the worst of all diseases, was a late complication of measles.

The information that McKeown collates in these pages is interesting not only for those with a taste for statistical figures but also for anyone drawn to social history. Thus it is especially interesting and rather shocking to learn how great a contribution infanticide has made to infant mortality; Disraeli, McKeown tells us, 'believed that infanticide "was hardly less prevalent in England than on the banks of the Ganges"'. McKeown reminds us, too, that both criminal and legal abortion are widespread and on the increase.

I agree with McKeown's assessment of the importance of infectious diseases in human mortality. I agree also with Haldane (whose name I do not see referred to in McKeown's book) that death from infectious disease is the most important selective force that has acted upon mankind and that it has left a very nearly indelible stamp on the human genetic constitution. To give one example only: the prevalence in West Africa of the gene which converts haemoglobin A into haemoglobin S seems to be owing to the fact that those who inherit this gene from one parent only enjoy a significant degree of protection against subtertian malaria. The gene does not cause major disability except when it is inherited from both parents, when it gives rise to the grave and usually fatal blood disease known as sickle cell anaemia, causing a loss of life statistically outweighed by the gain in protection from malaria. This is an instructive example because it is important evidence for the contention that improvement of the environment—the practice of 'euphenics' as President Joshua Lederberg of Rockefeller University calls it—can lead to genetic improvement (in this case the disappearance of gene S) rather than to genetic deterioration.

The causes of the great secular decline in mortality to which I have referred is one of the great problems of social medicine. In view of the nature of his thesis, it is not surprising that McKeown should quote with approval a passage from a presidential address to the American Association of Immunologists that attributes the secular improvement in health to the establishment of a new equilibrium between infectious organisms and their victims, 'quite regardless of our therapeutic efforts. According to this interpretation,' McKeown comments, 'the trend of mortality from infectious diseases was essentially independent of both medical intervention and the vast economic and social developments of the past three centuries.'

Professors of social medicine usually hold sewers in high esteem so it strikes me as surprising that McKeown evidently does not regard the institution of main sewage disposal systems as one of the 'social developments' to which he refers in the passage quoted just above. McKeown doesn't think much of antitoxins; nor, I must say, do I. I have however very little doubt of the efficacy of active immunization by toxoid substances, the evaluation of which is going to be complicated by the fact that today, persons specially at risk of contracting, say, tetanus are singled out for protective immunization and generally receive it if they have sensible and responsible physicians and employers. It is true, though, that cholera vaccine has not been proved to be efficacious. Polio vaccine, however, has been. As a medical scientist my own inclination is to give more weight to the success of the latter than the failure of the former but McKeown is determined to give the lion's share of the credit to the operation of natural selection: 'The immunological constitution of a generation is influenced largely by the mortality experience of those which precede it.'

I should be the last to depreciate the importance of natural selection and of evolutionary changes generally, but if they were a fully adequate safeguard against disease we should not get half the diseases we do. Dr David Pyke has shown, for example, that there is a clear-cut genetic element in susceptibility to the form of diabetes that presents itself in middle age or in older people. There is also a genetic element, though of a different kind, in differences of susceptibility to insulin-dependent diabetes of juvenile onset. The forces of natural selection working upon what was at one time a mortal complaint of early onset are immensely strong; but they have not been strong enough to eliminate the genetic constitution associated with a specially high susceptibility to insulin-dependent diabetes.

McKeown's views on the importance of nutrition in resistance to infectious disease are succinctly summarized as follows:

If the decline of mortality from infectious diseases was not due to a change in their character, and owed little to reduced exposure to micro-organisms before the second half of the nineteenth century or to immunization and therapy before the twentieth, the possibility that remains is that the response to infections was modified by an advance in man's health brought about by improved nutrition.

McKeown's case is founded upon the undoubted correlation between nutritional standing and susceptibility to infection, but since 'there is no direct evidence that nutrition improved in the eighteenth and early nineteenth centuries', we feel let down.

I shall use McKeown's own words to describe what he regards as evidence of improvement of food supply—his own words, lest in paraphrasing what he says I should be thought guilty of presenting an argument in such a way as to discredit him:

The most impressive evidence of the improvement in food supplies is . . . the fact that the expanded populations were fed essentially on home-grown food. The population of England and Wales increased from 5.5 million in 1702 to 8.9 in 1801 and 17.9 in 1851. Since exports and imports of food during this period were relatively small, it is clear that food production at least trebled to sustain an increase of 12.4 million in a century and a half.

The decline of mortality that occurred during the eighteenth and nineteenth centuries continued into the twentieth, but with the difference that in the twentieth century the reduction in mortality from non-infectious causes began to make an important contribution to the decline, particularly in respect of prematurity and diseases of early infancy. Deaths attributed to 'old age' diminished also, probably because improvements in diagnosis caused them now to be attributed to specific causes.

Infanticide, an important cause of death until at least the latter half of the nineteenth century, diminished during the twentieth, partly because the institution of foundling hospitals made it possible to dispose of children without killing them and partly because of the growth of contraceptive practices. The foundling hospital in St. Petersburg, McKeown reports, had 25,000 children in the mid 1830s on its rolls and admitted 5,000 annually; 30–40 per cent of the children died during the first six weeks and hardly a third reached the age of six. Those who denounce birth control procedures as morally the equivalent of murder might now pause to reflect that the reduction in the number of unwanted births has reduced the frequency of child murder—in a real, not figurative, sense.

McKeown summarizes the argument of the first and larger half of his book in terms which escape tautology only by a hair's breadth. The improvement of health that has taken place during the past three centuries was due

not to what happens when we are ill, but to the fact that we do not so often become ill; and we remain well, not because of specific measures such as vaccination and immunization, but because we enjoy a higher standard of nutrition and live in a healthier environment. In at least one important respect, reproduction, we also behave more responsibly.

Turning now to the future McKeown is simplistic to a degree that takes my breath away: 'there are only two ways in which disease occurs. It results either from errors in genetic programming at fertilization, or from . . . an environment for which the genes are not adapted.' To me, a biologist, this remark is about as illuminating as to be informed that disease is caused by a departure from a state of health.

These profundities usher in passages which Jean Jacques Rousseau would surely have applauded—passages in which McKeown says that whereas genetic adaptations in response to the impact of infectious diseases may occur 'within a few generations', the requirements for health of the digestive, cardiovascular, and reproductive systems do not differ greatly from those which prevailed during man's evolution—during which we were all nomadic and had practices in respect of diet and the expenditure of energy that were profoundly changed by the agricultural revolution and the accompanying domestication of man, and were of course still more greatly changed by the coming of industry. These passages pleased me because they are evidence that even an expert on social medicine still essentially falls in with the theory of illness that prevails throughout most of the Western world; I mean the 'punishment' theory of illness, according to which illness is a judgement upon us for indolence, sloth, gluttony, or other forms of carnal self-indulgence. These are salutary reflections that reaffirm the importance of the regulation of personal conduct. A new theory of illness is now taking shape at a time when the detritus of civilization is accumulating around us: the environment gets blamed for more and more that goes amiss and it is becoming increasingly easy to blame the environment or the iniquities of *laissez-faire* capitalism rather than, as in the old days, ourselves for our medical misadventures.

When he turns to considering our health in the future McKeown seems to me again to use too broad a brush for what is in any case too large a canvas. 'Most types of mental subnormality and of congenital malformations', he writes, are the consequence of prenatal environmental influences; that, surely, is too sweeping a statement.

We can only agree, though, that diseases of a kind which McKeown attributes to faulty genetic programming are relatively intractable, where diseases associated with affluence are in principle preventable. A miscellaneous group of diseases is classified as potentially preventable: 'some acute respiratory infections, such as the common cold, influenza and viral pneumonia as well as gastrointestinal diseases due to viruses. More tentatively, I suggest that many psychiatric conditions are in the same class.'

In a synoptic survey of the achievements of medicine McKeown makes the familiar and important point that the death-rate from tuberculosis underwent a progressive decline that was independent of the introduction of specific remedial measures. On the other hand he is inclined to dismiss as 'perverse' Creighton's view that vaccination played virtually no part in the decline of mortality from smallpox. Since he later expresses doubts on the efficacy of medical research it is heartening to see how clear is the evidence of the beneficial effects of vaccination against poliomyelitis.

It is fully in keeping with the character of McKeown's book that he doesn't think very much of medical research and that he should quote with delighted approbation Sir Macfarlane Burnet's extraordinary *lapsus mentis* in which he said that the contribution of laboratory science to medicine had come virtually to an end. The reason he took this view, I believe, is that Macfarlane Burnet was formerly, as I was, the head of a large medical research institute devoted to 'basic' medical research and that he was as dismayed as I was at the fact that so many members of his staff were more intent upon enlarging their own reputations as 'pure scientists' than in engaging directly upon the study of medical problems.

I now think that Burnet was quite wrong and that young scientists intent upon improving natural knowledge, and Lewis Thomas, who champions them, are right. As an antidote to Burnet's spiritless declaration I roundly declare that within the next ten years remedies will be found for multiple sclerosis, juvenile diabetes, and at least two forms of cancer at present considered somewhat intractable. These remedies, moreover, will come from medical research laboratories, very likely from people ostensibly working on some quite different subject.

It is one of the sadnesses of medical education that in spite of the earnest advocacy of people in the know, ordinary medical students tend to be bored by and are even a little contemptuous of the study of social medicine and public health. In British medical schools public health is traditionally taught alongside forensic medicine, in a course compendiously known to medical students as 'rape and drains'. I should now hazard an explanation of why so many medical students depreciate the importance of social medicine because it will help to explain why McKeown's book is likely to leave so many of its readers with a feeling of uneasy dissatisfaction. Social medicine, as McKeown expounds it in his book, has to do with the illnesses and mortality of whole populations and with how they vary from time to time and from place to place. On the other hand the feelings of compassion that are thought to tempt young students into medicine soon make them realize that it is individual people, not populations, that are ill and in need of treatment. For this reason I think it likely that medical education and medical research will for many years to come remain centred upon personal rather than social medicine.

16 Expectation and prediction

Wise folk may or may not form expectations about what the future holds in store but the foolish can be relied upon to predict with complete confidence that certain things will come about in the future or that others will not.

It is well worth while insisting upon the clear distinction of meaning between the two. A prediction always pretends to foreknowledge where an expectation is merely a hypothesis with a future setting ('I expect *Pluto's Republic* to be on sale in October 1982')—a hypothesis which the passage of time will either corroborate or confound. We cannot be viable human beings without taking some view of what will happen in the future. We confidently expect that the sun will rise tomorrow morning: it has become a habit of thought, but as the great Scottish philosopher David Hume (1711–76) pointed out we should be plunged into a labyrinth of philosophical difficulties if we attempted to *prove* that the sun would rise tomorrow on the basis that such a declaration would have been true every yesterday.

Astronomical predictions are perhaps the most famous of all and have in the past been the most awe-inspiring. They extend all the way from predictions such as one finds in nautical almanacs, giving us the exact times of sunrise, sunset, phases of the moon, high tides etc.; but grander than all of these, in the year 1704 the English astronomer and mathematician Edmund Halley predicted that the comet which now bears his name would return in 1758—it is again expected in this neck of the woods in 1986.

Surely such predictions embody foreknowledge? Not really: the future position of the comet is deduced from our *present* knowledge of the comet's whereabouts and of the nature of its orbit around the sun. Halley was in a specially good position to acquire this knowledge, for he was a friend of Isaac Newton's and thoroughly familiar with Newton's *Principia Mathematica*, widely thought of as the greatest of all scientific works, which he proof-read and saw through the press. Halley's spectacular display of apparent foreknowledge was based upon what he knew already of the comet's position plus a number of astronomical principles which he already knew, or thought or assumed he knew. If Halley had been mistaken it could only have been because the knowledge he thought he possessed in 1704 was mistaken, or his logical reasoning (that is, his mathematical calculation) was erroneous.

Astronomy and sociology: historicism

Halley's prediction made a tremendous impression and seemed to many people to be the very paradigm of all that is truly 'scientific'. It inflamed sociologists and historians with the ambition to devise a historical social science that would embody the laws of social transformation and of the historic process and thus make it possible to foretell the future of mankind—a truly scientific sociology, it was thought, would embody the laws of human destiny and so make historic prediction possible. The belief that a predictive science of history can exist is known as *historicism*.

To those familiar with such matters it is very well known that David Hume punctured the pretensions of the philosophic doctrine known as 'empiricism', the belief that upon the evidence of the senses—upon sensory information only—it is possible to propound scientific laws of apodictic certainty. Yet many who know of Hume's scepticism and the revolution he brought about in undermining empiricism are apparently unaware that the Vienna-born philosopher Karl Raimund Popper demolished historicism as effectively as Hume demolished empiricism.

Popper's argument turns upon the theorem that I shall attempt to derive formally below, namely that it is not possible to predict scientific ideas or advances in science. Popper's argument goes thus: the course of history is influenced—no matter how or how much—by advances of science and technology;[1] but advances in science and technology cannot be predicted: therefore the future course of history cannot be predicted.

An example of historicism: The economic interpretation of the process of history embodied in Marxism is the best-known form of historicist argument. I cite now an economic interpretation of history that illustrates very clearly the kind of fallacy to which Karl Popper called special attention. Consider the factors that determine the positioning, size and shape of factories and the whereabouts of the people who work in them. The source of energy being coal, which is costly to transport, factories must grow up alongside or near coalfields so as to reduce to the minimum the cost of transporting coal. The dwelling houses of workers will be near the factories, so a township must grow up in the neighbourhood of coalfields.

Coal is converted into usable energy through the medium of steam; it is not, however, feasible for each power-tool to be worked by its own little steam-engine. Factories will therefore be so arranged that a single big steam-engine can drive a single overhead shaft or spindle to which individual machine-tools are connected by flexible leather or rubber belts. Individual factories will get larger and larger so as to increase the number of machine-tools that can be served by a single steam-engine.

It all sounds perfectly plausible, but the argument is mistaken, for it is not true that factories get larger and larger and that they congregate nearer and nearer coalfields. Neither Marx nor any other devotee of the economic interpretation of history was able to foresee the coming of electrical power, one of the consequences of which was that factories no longer needed to be built near coalfields; moreover the properties of electrical power are such as to make possible the springing up of numerous small factories containing machine-tools individually powered by their own electrical outlets. This is just what has happened in the British Midlands and the State of Massachusetts—Marxists were wrong in the historical prediction based upon economic theory for the simple reason that Marx was unable to foresee the coming of electrical energy. A much more deeply erroneous prediction is that which turns upon the notion of class warfare. According to Marxist theory the direction of the flow of history is shaped by a struggle for supremacy between social classes—particularly between the proletariat and those who own the means of production. Marx predicted that the struggle between the classes would inevitably lead to a social revolution followed by the victory of the proletariat and the disappearance of class stratification. Because of the paramount necessity to maximize profits, the rate of pay, degree of freedom and general welfare of the working class must inevitably deteriorate—and this would be a principal factor in bringing about that social revolution towards which history was inevitably proceeding.

These predictions have not been fulfilled: so far from deteriorating progressively the lot of the working classes in the great industrial countries in which their welfare was thought to be most gravely at risk has slowly and progressively improved.

The other great historicist doctrine that has dominated the political thought of the twentieth century is Fascism. The high Fascism of Nazi apologists such as Alfred Rosenberg is a kind of racial or genetic élitism which declares that social progress and the advancement of mankind is rather specially the privilege and responsibility of a racial élite whose inborn superiority is seen to best advantage in the German peoples, who by reason of this inborn superiority will of historic necessity conquer the lesser nations and thus rise to their full stature as rulers of the world.

We still have reason to thank God that this transformation of the world never came about and that the doctrine of genetic élitism is now for the most part confined to morally criminal sects in Britain and the USA. It is not thought likely that it will ever again become a major factor in the causation of wars on a global scale.

It would be a mistake to think that historicism considered as a cultural disease is no longer a threat, for there are several contexts in which historicist thinking still flourishes. The most important of these are in economics and demography, which I shall deal with in turn.

Economic prediction and prediction of the weather

There are certain striking similarities of principle between economic prediction and weather prediction—apart, that is, from the fact that both may be comically wrong. There are important differences too—and these are such as to make prediction of the weather easier to execute and generally speaking more reliable than economic prediction.

The similarities are that both deal with multivariate systems of enormous complexity and of great intrinsic difficulty (a multivariate system is one whose properties are determined by the values of not just one or two but a great number of factors that vary independently and are individually subject to sampling errors and errors of ascertainment). In both the prediction is deductive in style, that is to say the prediction is about the future state of a system of which the present state is known or assumed to be known and the pattern of interaction between the variable quantities is also assumed to be known. The differences are set out in the accompanying table, which I now explain.

Economic and weather prediction compared
(see text for explanation of table)

Weather	*Economy*
Variables scalar	often non-scalar
Functional relations often exactly known	usually not known
Uninfluenced by politics or fashion	influenced by both
Wholly non-reflexive	highly reflexive

1 Weather prediction has to do with scalar variables, that is to say, simply measurable or numerable quantities such as inches of rainfall, wind speeds, per cent cloud cover, humidity, barometric pressure and so on. But many important non-scalar variables enter into economic prediction: confidence, for example. No single factor has a greater influence upon the economy, but how is it

to be measured, if at all? Again, economic prediction often depends upon political predictions that are notorious for their fallibility ('In my opinion the Democrats will sweep back into power').

2 In weather prediction the functional relations between variables are physical in character and have been ascertained from long-standing empirical experience—I mean the relations between temperature, barometric pressure, humidity and the likelihood of precipitation. But with the economy comparable functional relations are mostly conjectural in character; their equivalent in the world of meteorology would be propositions such as 'When the wind blows, the cradle will rock.'

3 'Fashion' is almost by definition capricious and unpredictable. The weather is wholly immune to the vagaries of fashion while the economy is not.

There are special circumstances in which political decisions might influence the weather—for instance, a political decision increasing the number of trials of nuclear weapons. But being themselves unpredictable such considerations do not enter meteorological forecasts; they do, however, enter into the economic forecasts. They are subject to two sources of error: the political presumption may be mistaken and so also may be the nature of the political influence on the economy. Thus the coming into power of a strong right-wing government in Boolooland may strengthen the zlotnik (Bernard Levin's generic term for all units of foreign currency) but the new president's aggressive attitude towards his neighbours may arouse alarm and thus cause the value of the zlotnik to tumble.

4 'Reflexive' is a technical term used in logic and grammar and meaning 'referring to or acting upon itself'. To describe weather prediction as 'non-reflexive' is merely to say, what is self-evidently true, that weather predictions do not affect the weather. It is notorious, though, that economic predictions can affect the economy: economic predictions are strongly reflexive in character. For this reason especially, though there are many others, weather prediction is sounder in principle than economic prediction—is altogether a safer bet. Another circumstance that makes

weather prediction easier in principle than economic prediction is the apparent stability of the factors that influence climate, which means that meteorologists can appeal to precedents and historical evidence dating back as far as records exist—for after all (I quote D'Arcy Thompson) 'A snowflake is the same today as when the first snows fell.' It is far otherwise with the economy because the factors that influence the economy change not merely from year to year but even from day to day (think of Middle Eastern oil, for example). In short, the economy has changed far more radically since the days of Adam Smith than the weather has since Benjamin Franklin's.

It amazes me therefore that people who are far too canny and experienced to take weather predictions as certainties or to make their plans and perhaps alter their life-styles in accordance with them are yet almost completely gullible when it comes to economic prediction. Since they are of great intrinsic difficulty, beset by all the special difficulties that have just been outlined, we cannot wonder that economic predictions are often grievously mistaken.

It is not their wrongness so much as their pretensions to rightness that have brought economic predictions and the theory that underlies them into well-deserved contempt. The dogmatic self-assurance and the asseverative confidence of economists are additional causes of grievance—self-defeating traits among people eager to pass for scientists.

Demographic prediction

Demography is the branch of sociology that has to do with populations: their structure, distribution and reproduction. Short-term demographic predictions are necessary, desirable, and in the main not grossly inaccurate. Knowing the number of births today and assuming that there will be no dramatic change in mortality during the next five or six years, we can predict with adequate accuracy how many places we should provide in primary schools five years hence, or in high schools ten to twelve years hence. This is straightforward enough, but long-term predictions are especially prone to error.

The high tide of population prophecy was in the 1930s when fertility in the countries of northern Europe and the USA had fallen so low as to raise in many minds the question of whether the peoples of the Western world might not die out altogether through non-replacement. Books were written with titles such as *The Twilight of Parenthood* (by Dr Enid Charles), and it is a measure of the gravity of the situation that the world's foremost demographer, the former assistant statistician of the Metropolitan Life Assurance Company, was moved to poetry in commenting a few years later on the situation.

It surely would be a strange trick of fate [he remarked] if we, the most advanced product of organic evolution, should be the first of all living species so clever as to foresee its own doom.
> Like some bold seër in a trance
> Seeing all his own mischance—

We have eaten of the fruit of the Tree of Knowledge, and it has turned to poison.

So wide of the mark were these gloomy prognostications that it was not long before the fear of depopulation was supplanted by our current principal cause of concern: over-population. Human beings are so riotously fertile, it is thought, and so effectively do medicine and hygiene now protect them from infectious disease and other major causes of mortality, that mankind faces a threat of over-population so severe as to put us at the mercy of starvation and new pestilences of unimaginable severity.

The threat of over-population should on no account be underestimated, but it must be remembered that the predictions which give rise to it do not have the apodictic certainty that was at one time imputed to them, for new and unforeseeable factors have been found to enter into the equations of mortality and fertility, one of them being that an increase in the standard of living may be accompanied by a diminution of fertility rather than by the still greater increase that had at one time been feared. We are by no means out of danger yet but I feel that the period of acute panic is over and has been giving way to a more sober assessment of the practicality and effectiveness of the procedures that might help to keep populations within reasonable bounds.

Learning from past mistakes, as economists sometimes do not, demographic prediction has improved greatly and commands far greater respect now than hitherto. One of the great mistakes made in the past was to suppose that a population's reproductive well-being and the probability of its maintaining its numbers could be measured by a single figure such as the net reproduction ratio or 'Malthusian parameter', much as a patient's degree of fever can be represented by a single thermometer reading—a procedure that has no parallel in demography.[2] Demographic predictions are now based upon the assessment of variables that have a real meaning in terms of the way people actually behave: variables such as marriage rates, marriage ages, ages in life at which children are born—that is, they include the pattern of family building etc. Above all, the *cohort* tends to have supplanted the population as a whole or some sub-category of it as the subject of prediction. A 'cohort' consists, for example, of all the people born in one year or married in one year or going to school in one year, and the fertility and mortality of each cohort can be followed throughout life until they all drop out. This is sometimes spoken of as a 'longitudinal' as opposed to a cross-sectional way of looking at the population.

Improved thought they may be I do not think that we shall ever be able to rely on demographic predictions as we rely upon astronomical predictions, but so much of educational, fiscal, military and political importance turns upon population numbers that we must always try: there can be no abdication from this responsibility, however imperfect the process of prediction may be; but these imperfections are such that we shall probably never be wholly free from fear of over-population or of some degree of depopulation.

Negative predictions

No kind of prediction is more obviously mistaken or more dramatically falsified than that which declares that something which is possible in principle (that is, which does not flout some established scientific law) will never or can never happen. I shall choose now from my own subject, medical science, a bouquet of negative predictions chosen not so much for their absurdity as for the way in which they illustrate something interesting in the history of science or medicine.

My favourite prediction of this kind was made by J. S. Haldane (the distinguished physiologist father of the geneticist J. B. S. Haldane), who in a book published in 1930 titled *The Philosophy of Biology* declared it to be 'inconceivable' that there should exist a chemical compound having exactly the properties since shown to be possessed by deoxyribonucleic acid (DNA). DNA is the giant molecule that encodes the genetic message which passes from one generation to the next—the message that prescribes how development is to proceed. The famous paper in the scientific journal *Nature* in which Francis Crick and James Watson described the structure of DNA and how that structure qualifies it to fulfil its genetic functions was published not so many years after Haldane's unlucky prediction.[3] The possibility that such a compound as DNA might exist had been clearly envisaged by the German nature-philosopher Richard Semon in a book *The Mneme*, a reading of which prompted Haldane to dismiss the whole idea as nonsense.

In the days before the introduction of antisepsis and asepsis, wound infection was so regular and so grave an accompaniment of surgical operations that we can hardly wonder at the declaration of a well-known surgeon working in London, Sir John Erichsen (1818–96), that 'The abdomen is forever shut from the intrusions of the wise and humane surgeon.' Of course, the coming of aseptic surgery to which I refer below, combined with the improvement of anaesthesia, soon made nonsense of this and opened the door to the great achievements of gastrointestinal surgery in the first decade of our century.

One of the very greatest surgeons of this period was Berkeley George Moynihan of Leeds (1865–1936), a man whose track-record for erroneous predictions puts him in a class entirely by himself.

Around 1900 the famous British periodical, the *Strand* magazine (the first to publish the case records of Sherlock Holmes), thought that at the turn of the century its readers would be interested to know what was in store for them in the century to come; 'a Harley Street surgeon' (unmistakably Moynihan) was accordingly invited to tell them what the future of surgery was to be. Evidently not spectacular, for Moynihan opined that surgery had reached its zenith and that no great advances were to be looked for in the future—nothing as dramatic, for example, as the opening of the abdomen, an event regarded with as much awe as the opening of Japan.

Moynihan's forecast was not the hasty, ill-considered opinion of a busy man: it represented a firmly held conviction. In a Leeds University Medical School magazine in 1930 he wrote: 'We can surely never hope to see the craft of surgery made much more perfect than it is today. We are at the end of a chapter.' Moynihan repeated this almost word for word when he delivered Oxford University's most prestigious lecture, the Romanes Lecture, in 1932. He was a vain and arrogant man, and if these quotations are anything to go by a rather silly one too, but surgery is indebted to him nevertheless, for he introduced the delicacy and fastidious-ness of technique that did away for ever with the image of the surgeon as a brusque, over-confident and rough-and-ready saw-bones. Moreover Moynihan, along with William Stewart Halsted of Johns Hopkins (1852–1922), introduced into modern surgery the *aseptic* technique with all the rituals and drills that go with it: the scrupulous scrub-up, the gown, cap and rubber gloves, and the facial mask over the top of which the pretty young theatre nurse gazes with smouldering eyes at the handsome young intern who is planning to wrong her. These innovations may be said to have made possible the hospital soap opera and thus in turn TV itself—for what would TV be without the hospital drama, and what would the hospital drama be without cap and masks and those long, meaningful stares?

The full regalia of the surgical operation did not escape a certain amount of gentle ridicule—in which we may hear the voice of those older, coarser surgeons whom Moynihan supplanted. Moynihan was once described as 'the pyloric pierrot', and upon seeing Moynihan's rubber shoes a French surgeon is said to have remarked 'Surely he does not intend to stand in the abdomen?'

It might easily be thought that the errors of judgement to which I have called attention are the work of individual men and that committees would be sure to do better. It would be wishful thinking to entertain any such belief. In a report published at the end of the 1914–18 war the leading administrative body of medical high-ups in Great Britain wrote: 'a paraplegic patient may live for a few years in a state of more or less ill-health'.

This remark was quoted a little sadly by the neurologist Ludwig Guttmann in the introduction to his great text on injuries of the spinal cord, *Spinal Cord Injuries: Comprehensive Management and Research*—a work written while he was planning the first paraplegic Olympic Games in Israel.

In the appraisal of all these misjudgements we must not forget that it is the erroneous prediction that tends to stick in the mind: there is not the same incentive to remember the predictions that are right, because if they are self-evidently true the likelihood is that they are also rather dull. Thus the great Canadian physician Sir William Osler always insisted that chemistry would play an increasingly important part in modern medicine—and he could hardly have been more right.

Nevertheless, some medical high-ups do seem to be rather especially prone to making confident predictions. I believe this to be due to the steeply hierarchical nature of the medical profession, which makes the high-ups so accustomed to the deference and even sycophancy of medical students, nurses and their juniors generally as to engender in them the idea that their opinions are especially likely to be right because it is they who hold them.

The philosophy of prediction

In the discussion of astronomical predictions—among which I include all predictions that turn on logical deductions from premises believed, hoped or assumed to be true—there is no element of foreknowledge at all: the prediction is based upon *present* knowledge which of course may or may not be right.

I now propose to show that the most interesting—and what would be the most valuable—kind of prediction, the prediction of future ideas, is not logically possible, and that it is a fallacy to suppose that although long-term predictions are highly fallible, short-term predictions can be exactly right.

The arguments I shall bring forward turn upon simple logical paradoxes, the evidence of which is conclusive, for any line of reasoning that embodies a self-contradiction must certainly be mistaken. On no account belittle the paradox, therefore, or dismiss it as a mere play upon words: a paradox has the same significance for the logician as the smell of burning rubber has for the electronics engineer.

First consider the self-contradictoriness of supposing that one can predict future ideas ('future' means a day, a week, a year or even ten minutes hence). Consider the following statement: 'I predict that in 1983 research workers at New York's Memorial Sloan-Kettering Cancer Center will propound the entirely novel theory of the role of viruses in the causation of cancer, to wit . . .'. There are now only two possibilities. The statement can be completed or it can be left incomplete. If the statement is completed the theory whose promulgation I predict cannot be wholly new in 1983 because to complete the sentence I shall have to propound it here and now, so it would not be a future idea at all; if on the other hand I leave the declaration incomplete then I am not making a prediction. (It was considerations such as these that led Karl Popper to refute the pretensions of the cultural disease of modern sociology that he described as *historicism*.)

It might be maintained that whereas long-term predictions are altogether too hazardous for us to have any confidence in them, nevertheless short-term predictions can sometimes be dead accurate. But this leads to a paradox too, for accurate short-term prediction implies accurate long-term prediction, which we have just conceded to be impossible. Consider for example a prediction made in 1982 on any topic whatsoever relating to the year 1990, and let us agree that it is impossible that so distant a prediction should be accurate. If, however, we could predict accurately all that will happen as soon as next year, this prediction will include an accurate statement of what someone in 1982 is going to say about what will happen in 1983; this in turn will include what someone in 1983 is going to say about what will happen in 1984, and so on until we reach 1990, and we shall end by having made the accurate long-term prediction that we have declared to be impossible. Accurate short-term prediction logically entails accurate long-term prediction—a palpable self-contradiction.

Advice to a young reader

My own interest in prediction grew out of the series of articles in the *Strand* magazine to which I referred above, a series in which various eminent people were invited at the turn of the century to forecast what would happen in the following century.

The turn of the millennium (1999/2000) will make severe demands upon soothsayers and I expect some readers of this article will be invited to make pronouncements upon, for example, 'air travel/psychiatry/podiatry in the third millennium AD'. I should like to give literally one word of advice to all who are invited to make such predictions: *Don't*.

17 Scientific fraud

Some policemen are venal; some judges take bribes and deliver verdicts accordingly; there are secret diabolists among men in holy orders and among vice-chancellors are many who believe that most students enjoying higher education would be better-off as gardeners or in the mines; moreover, some scientists fiddle their results or distort the truth for their own benefit.

None of these, though, is representative of his profession—and only people young enough to be cynical believe them to be so. The number of dishonest scientists cannot, of course, be known, but even if they were common enough to justify scary talk of 'tips of icebergs' they have not been so numerous as to prevent science's having become the most successful enterprise (in terms of the fulfilment of declared ambitions) that human beings have ever engaged upon. The profession, sticking together (which is not such a bad thing to do), believes that cheating in science is a curious minor neurosis like cheating at patience—something done to bolster up one's self-esteem. Rather than marvel at, and pull long faces about, the frauds in science that have been un-covered, we should perhaps marvel at the prevalence of, and the importance nowadays attached to, telling the truth—which is something of an innovation in cultural history, if by the truth we mean correspondence with empirical reality. The authors of the more lurid travellers' tales would have been taken aback if some-one had described them in modern vernacular as 'bloody liars', but so they were, many of them. They were telling stories, and wanted to tell good stories. Aristotle's conception of poetic truth was one in which correspondence with reality played little part, and his biology gave an account of what he thought *ought* to be true in the light of his deep conception of the true purposes of nature. Thus it ought to be true according to the hebdomadal rule that male semen is infertile between the ages of seven and twenty-one—a pathetic absurdity of which Aristotle would not have been guilty if he had had any real sense of empirical truth. Aristotle was a pioneer, perhaps, in what I believe to be the commonest form of self-deception in science: the kind of attachment to a dearly loved hypothesis that predisposes us (yes, all of us) to attach a special weight to observations that square with and thus uphold our pet hypotheses, while finding reasons for disregarding or attaching little weight to observations and experiments that cast doubt upon them. There is no one who does not roll out the welcome mat with a flourish for those who bring evidence that upholds our favourite preconceptions.

The most puzzling fraud of all—for such in effect it was—was that of the segregation ratios (3:1; 9:3:3:1) as reported by Gregor Mendel in his plant-breeding experiments. As R. A. Fisher was the first to point out, these ratios conformed far too closely to theoretical expectations to be plausible, having regard to the numbers of plants and seeds involved. The explanation could be as simple as that Mendel was a nice chap whom his gardeners and other assistants wanted very much to please, by telling him the answers which they suspected he would dearly like to hear: moreover, as Mendel was an abbé, his assistants may have felt that there was an element of heresy in securing results other than those the Reverend Father was convinced were true. This is a subject on which the authors of the present book[1] write amusingly.

I do not suppose that personal advancement is a principal motive for cheating in science: rather it is the hunger for scientific reputation and the esteem of colleagues. And I believe that the most important incentive to scientific fraud is a passionate belief in the truth and significance of a theory or hypothesis which is disregarded or frankly not believed by the majority of scientists—colleagues who must accordingly be shocked into recognition of what the offending scientist believes to be a self-evident truth.

Two scientific theories or viewpoints are notorious for arousing this passion: the doctrine of the inheritance of acquired characters associated with the name of J.-B. P. A. de Monet, le Chevalier de Lamarck, on the one hand, and the farrago of sillinesses that may be compendiously called 'the IQ nonsense', on the other. Consider Lamarckism first. One kind of Lamarckian inheritance is so commonplace and obvious as to be recognized for what it is by anyone who gives the matter a thought: it is that in which parents or members of a parental generation impart to their children or in general to a filial generation the knowledge and skills they had themselves acquired during their lifetimes. This is heredity all right, but it is 'exogenetic' in character, in the sense that it is not mediated through the genetic plant of chromosomes and genes, but through precept, example, and deliberate indoctrination. Unlike ordinary or endogenetic heredity, this other kind is reversible and is Lamarckian in style, for that which is acquired in one generation may be transmitted to the next and so on, cumulatively. The existence of this mode of heredity has given people a powerful incentive to believe that ordinary or genetic heredity works in this way too, as it seems only natural justice that it should, and even professional biologists have been taken in by the parallel between exo- and endo-genetic heredity and by what looks like a constitutional inability to realize this is not how nature works. The mechanism of heredity is selective, not instructive: what happens in an organism's lifetime, even if it is a profound bodily modification brought about by an adaptive response, cannot be imprinted upon the genome. There is no known or even conceivable genetic process by which DNA can be taught anything. It seems most unjust that this should be so, but so it is, for in heredity a person's exertions to improve his body or mind to adapt himself to new environments all go for nothing.

Lamarckian inheritance is a topic upon which literary people have for some reason felt themselves entitled to express an opinion. It is entirely understandable that George Bernard Shaw should have done so, but less obvious why Samuel Butler should have been among their number, especially as he expressed better than anyone else the essence of the teaching of that August Weismann who overthrew Lamarckism. Butler said that according to Weismann 'a hen is simply an egg's way of making another egg'. Lamarckism has on at least one occasion been the subject of a scientific fraud, as recounted in Arthur Koestler's *Midwife Toad*. Koestler disclaimed being a Lamarckist, but created an atmosphere favourable to Lamarckism by representing Darwinian geneticists as the spokesmen for a dull and unperceptive establishment of conventional belief. Biologists and psychologists who have been won over to Lamarckian thinking have published a whole number of experiments which purport to demonstrate Lamarckian inheritance, but all such demonstrations have been faulty. Either they have not excluded an orthodox Darwinian interpretation, or they have been open to explanations of other kinds, or they have been technically faulty. I myself was involved as a spectator in one such attempt. There was no dishonesty, indeed nothing more culpable than self-deception—in this case, the enthusiastic selection, from results that were all over the place, of only those that fitted the hypothesis the experimenter was seeking to corroborate.

Why should Lamarckism arouse such passionate conviction of its truth? I believe that the well-known association of Lamarckism with the sinister and indeed evil opinions of Trofim Denisovich Lysenko points to a political explanation. Lamarckism seems only fair: is it not right that mankind should benefit from their exertions and utterly wrong that man's genetic provenance—his breeding, in fact—should determine absolutely his character, capabilities, and deserts? It was his well-founded suspicion that his teachings tended to question the pre-eminence of a man's breeding that caused Napoleon's contemptuously dismissive attitude towards Lamarck. To a man convinced that Lamarckian inheritance is true because it is fair and socially just, it seems that selectionist theory presents an attempt, in Condorcet's words, to 'render nature herself an accomplice in the crime of political inequality'.

The second of the two major causes of that passionate belief that can conduce to fabrication was that which I referred to as 'the IQ nonsense', by which I mean the complex of beliefs arising out of a contention of H. J. Eysenck's which I lose no opportunity to hold up to public ridicule. 'Clearly the whole course of development of a child's intellectual capabilities is largely laid down genetically . . .' (*The Inequality of Man* (1973), p. 111).

The most shocking deception arising out of a passionate acceptance of the idea that intelligence is susceptible to a scalar measurement and is 90 per cent heritable was the lengthy and studied scientific frauds of Sir Cyril Burt in his measurement of the IQs of twins reared together or separated from birth. This fraud was uncovered by Dr Leon Kamin and a skilful geneticist turned investigative journalist, Dr Oliver J. Gillie. In the present book, which gives a lot of attention to this case, Broad and Wade illustrate the inefficacy of scientific monitoring within the profession itself—of the procedures which those of us who maintain the integrity of scientists believe prevent or rectify scientific fraud. But the reason Burt's findings were not subjected to intent and independent critical scrutiny is simple and understandable. There was no effective check of Burt's findings because he told the IQ boys exactly what they wanted to hear. The fault lay not with the scientific monitoring system but with the bigotry and deep-seated misconceptions of the champions of the IQ concept.

The present authors greatly enlarge our understanding of the Burt frauds by recounting how a graduate student of Iowa State University, Leroy Wolins, 'wrote to 37 authors of papers published in psychology journals asking for the raw data on which the papers were based'. No fewer than 28 reported that their data had been misplaced, lost, or inadvertently destroyed.

The difficulty of laying hands on the 28 sets of data that were 'lost' or withheld was made somewhat more comprehensible by the horrors that emerged from the nine sets made available. Of the seven that arrived in time to be analysed, three contained 'gross errors' in their statistics. The implications of the Wolins study are almost too awesome to digest. Fewer than one in four scientists were willing to provide their raw data on request, without self-serving conditions, and nearly half of the studies analysed had gross errors in their statistics alone. This is not the behaviour of a rational, self-correcting, self-policing community of scholars.

Burt's is only one of the many notorious cases of fraud the authors deal with. All the old favourites are to be found in the index: Piltdown, Paul Kammerer of *The Midwife Toad*, and the infamous William T. Summerlin.[2]

A shocking story. Yet it is not the authors' intention to shock, though in fact they do so; no, the purpose is rather to show that research is not a wholly rational and explicitly logical procedure but subject to the confinement and constraints that afflict other professional men trying to make their way in the world. Moreover it questions our comfortable assumption that scientific cheating is very rare—an exceptional event that does not become a serious threat because science is protected by a whole number of built-in professional safeguards which bring it about that fraud is soon uncovered and the culprit punished.

The authors are very experienced professional science writers, and have made a highly responsible and well-argued contribution to the sociology of science. In spite of these sterling virtues, their book contrives also to be interesting and readable, and suitable for a lay readership. Even science writers are sometimes frauds, and the present authors, though they must have been aware of it, make no mention of a book by a science writer giving an account of a professedly authentic example of human cloning, ingeniously tricked out with quasi-scientific references put in 'to add corroborative detail to an otherwise bald and uncovincing narrative'.

What lesson should the scientific profession learn? Should we henceforth go around on our guard, doubting and questioning, looking for fraud and misrepresentation with the air of men expecting to find evidence of it? No, indeed not. Listening for a second time to Sir Kenneth Clark's splendid series of television broadcasts on 'Civilisation', I was again struck by the importance that Clark attached to confidence as a bonding agent in the advance of civilization, as it is indeed throughout professional life. Do not lawyers, bankers, clergymen, librarians, and editors tend to believe their fellow professionals unless they have a very good reason to do otherwise? Scientists are the same. The critical scrutiny of all scientific findings—perhaps especially one's own—is an unqualified desideratum of scientific progress. Without it science would surely founder—though not more rapidly, perhaps, than it would if the great collaborative expertise of science were to be subjected to an atmosphere of wary and suspicious disbelief.

18 Son of stroke[1]

Where to be ill. Large teaching hospitals are recommended. Unless privacy is of overriding importance or you really dislike your fellow men don't go into a private ward. The nursing won't be better than in a public ward, and may easily be much worse. Besides, in a public ward you will be entertained all day by the unfolding of the human comedy and by contemplating what literary people call the Rich Tapestry of Life.

Long stays in hospital. Lying in bed for any length of time is itself a weakening process, as you will soon find when you try to get up. In adequately staffed hospitals, however, physiotherapists will keep your muscles and joints in working order.

An analogous treatment is necessary for the mind. It is a natural tendency of the mind to come to and remain at a complete standstill. This is a principle of Newtonian stature. Prolonged disuse of the brain is also bad for you. Try therefore to think or converse about something other than the exigencies of hospital life and your own piteous plight. Guests come in useful here (see below: *Visitors*) and so do books.

Books. Books, if you are well enough to read them, are crucially important for entertainment and keeping the mind in working order. Some serious works should therefore be among them. Remember, however, that if you didn't understand Chomsky when you were well, there is nothing about illness that can give you an insight into the working of his mind. Do not read a genuinely funny book within a week of having had an abdominal operation. So far from giving you stitches, it will probably deprive

you of them. Books should never be so heavy as to impede the
ebb and flow of the blood. A slender anthology of selected English
aphorisms is strongly recommended. Ten aphorisms are normally
reckoned to be equivalent to a quarter of a grain of phenobarbi-
tone. Never take more than twenty aphorisms without medical
supervision.

Sleep. If you sleep all day you must not be aggrieved if you don't
sleep all night. If wakeful don't clamour for sleeping draughts, but
take ten selected English aphorisms with a cup of warm milk (see
above: *Books*).

Food. The food in hospitals is surprisingly good, but was not
intended for people with dainty or fastidious appetites. Be warned
that if you eat all day you will become disgustingly obese and thus
very properly an object of derision to your friends. Desist, therefore, and give those chocolates to the nurses.

Radio. It is traditional for hospital beds to be equipped with radio
outlets that don't work. Test the radio at the earliest possible
opportunity, complain as soon as possible, and go on complaining
until somebody does something about it. When the radio works
see that kind friends bring in the *Radio Times*. Then you won't
reproach yourself for missing that talk on the vegetation of
Boolooland. Small transistor radios are fine, provided they have
an ear monophone attachment. Otherwise they are as offensive as
smoking pipes.

Sister. Your ward sister is well worth knowing and trying to make
friends with, because she is almost certain to be an unusually
capable and intelligent woman, which is just as well because she
is a nurse, teacher, administrator, psychotherapist, and everybody's confidante. You are doing well if you manage to make
friends with her.

Nurses. The qualities of character which induce young ladies to
enter this overworked and underpaid profession are such as to
make them specially likeable people. You will almost certainly
want to do something to show your appreciation of them. Flowers
and profuse gratitude are not very imaginative. It is a fact, how-

ever, that nurses are often ravenously hungry after a day's duty
on the wards or soon after coming on duty after a characteristically inadequate breakfast. A secret supply of biscuits and cheese
may be more acceptable and will certainly be more digestible than
a pot of hothouse blooms. Another trait which nurses find agree-
able is to make sure that you are visited by a stream of handsome
and preferably unmarried sons, cousins, or brothers.

Visitors. Some visitors come because they love you or are genuinely concerned for you, and these you will generally welcome.
Others come because they felt they ought to or to indulge their
Schadenfreude. The latter should be got rid of as quickly as possible.
This can only be done by prior arrangement with Sister who is
adept at making unwanted visitors feel, as well as merely being,
unwelcome.

The bodily motions. In some wards the nursing staff give the impression of regarding it as a personal affront if the entire mucosal
lining of the great bowel is not evacuated daily. They attach much
more importance to this than you need.

It is a rightly humiliating thought that, in spite of Man's ability
to reach the moon, etc., etc., no one has yet desgined a bedpan
which is not physiologically inept, uncomfortable, and somewhat
obscene. The main factor in making physiotherapy supportable is
the feeling that ultimately it will equip you to get out of bed
yourself and look after your own needs.

Hospitality to guests: drinking. It has been said that the Middlesex
Hospital will do anything for you except allow you to park in the
forecourt, and in general the great teaching hospitals were erected
at least half a mile from anywhere it is possible to park a car. This
means that your visitors when they arrive will be harassed and
exhausted and must be offered the drink which (if they have any
sense) they will have brought with them. They will probably offer
you a drink at the same time, but as the words 'Thanks, I don't
mind if I do' rise to your lips remember the medical staff may
easily mind quite a lot. They certainly will if you are suffering
from a serious liver disorder. If your complaints are merely ortho-

paedic or mental they are not likely to object at all. But here again consult with the ward sister. Tell her, if need be, that you get a funny sort of dizzy swimming feeling in the head if you don't have a drink at six o'clock.

Serious illness: the will to live. A well-known public figure who has taken it upon himself to become the Conscience of the World has objected to transplantation as an unnatural and somewhat unwholesome method of prolonging life. But before they insist too vehemently upon the Right to Die, such people should remember that a very decided preference for remaining alive has been a major motive force with human, as with animal evolution. A very firm determination to remain alive has a mysterious therapeutic effect which helps to promote that very ambition.

The National Health Service. Don't run down the National Health Service which, in spite of faults which are inevitable in any manmade scheme, represents the most enlightened piece of social legislation of the past hundred and fifty years. If you think you can do better as a private patient attending private clinics, then good luck to you. You may need it.

19 The question of the existence of God

Because of the especially important place he occupies in the philosophy of science, I gave Francis Bacon his say near the beginning of this essay.[1] Now, at about the same distance from the end, I think he should be allowed to speak again. Francis Bacon was a simply reverent man, in spite of the traits in his thought that led Paolo Rossi to describe him as 'a medieval philosopher haunted by a modern dream'.[2] In his Confession of Faith, Bacon wrote thus: 'I believe that nothing is without beginning but God; and no nature, no matter, no spirit, but one only and the same God, that God as He is eternally almighty, only wise, only good in His nature, and so He is eternally Father, Son and Spirit in persons.'

As the result of some spiritual blindness or deficiency disease, I do not share Bacon's simple reverence, though I know that his belief in God is very widely shared. On the contrary, I believe that a reasonable case can be made for saying, not that we believe in God because He exists but rather that He exists because we believe in Him. In spite of the suspicion that rightly attaches to epigrammatic declarations that are tainted by smartness, the element of truth in the argument I propose to propound has long been recognized in such familiar and flippant blasphemies as 'Man created God in his own image.'

God and Popper's Third World

In *Pluto's Republic*, I summarized and explained Karl Popper's conception of a third world, inhabited by the creations of the mind, in the following terms:

Human beings, Popper says, inhabit or interact with three quite distinct worlds: World 1 is the ordinary physical world, or world of physical states; World 2 is the mental world, or world of mental states; the 'third world' (you can see why he now prefers to call it World 3) is the world of actual or possible objects of thought—the world of concepts, ideas, theories, theorems, arguments and explanations—the world, let us say, of all the furniture of mind. The elements of this world interact with each other much as the ordinary objects of the material world do: two theories interact and lead to the formulation of a third; Wagner's music influenced Strauss's and his in turn all music written since. Again, I mention that we speak of things of the mind in a revealingly objective way: we 'see' an argument, 'grasp' an idea, and 'handle' numbers expertly or inexpertly as the case may be. The existence of World 3, inseparably bound up with human language, is the most distinctively human of all our possessions. This third world is not a fiction, Popper insists, but exists 'in reality'. It is a product of the human mind but yet is in large measure autonomous.

In his own account of his idea (*Objective Knowledge: An Evolutionary Approach*, Oxford, 1972), Popper gives more emphasis than I do to the third world's containing theoretical systems, arguments and problem situations.

Considered as an element of this third world, God has the same degree and kind of objective reality as do the other products of mind. It goes with believing in God that we address Him with praise and reverence and obey Him or are otherwise influenced by Him; we make images of Him and believe ourselves to be made in His image. In prayer, we enter into imaginary dialogue with Him and seek from Him comfort and advice. Finally, we believe in God as an agent—indeed, as Prime Mover. God's objective existence rests upon our belief in Him; if that belief were to cease, the reverence and the dialogue would end and we should no longer look to Him as Prime Mover.

Where so many people I like and admire do so and derive strength and comfort from doing so, I am not at all proud of my lack of belief, and while it would not be in my power to simulate belief (a deception that would soon be unmasked), I should like my behaviour—short of overt acts of worship or the avowal of beliefs I do not hold—to be such that people take me for a religious man in respect of helpfulness, considerateness and other evidences of an inclination to make the world work better than it otherwise would be. In short, I should like to be thought to possess what it rightly enrages Jewish people to hear described as the 'Christian virtues'.

I regret my disbelief in God and religious answers generally, for I believe it would give satisfaction and comfort to many in need of it if it were possible to discover and propound good scientific and philosophic reasons to believe in God.

It would not be just to attribute my disbelief to my having led a sheltered academic life less exigently in need of comfort and support than those whose lives have been turbulent or unhappy or in other ways more at risk than my own. Twice in my life I very nearly died as a result of cerebral vascular accidents, and I don't look forward a bit to making, in due course, a clean job of it. I neither cursed God for depriving me of the use of two limbs nor thanked and praised Him for sparing me the use of two others. On these two occasions I derived no comfort from religion or from the thought that God was looking after me; indeed, if I had not disapproved of his famous poem on literary grounds—this kind of braggadocio is a pain—I should have derived more comfort from the William Ernest Henley, who professed to be master of his fate and thanked God for his unconquerable soul. But there was no comfort here. No one knew better than I did that no one is undefeatable: that's just heroics. What matters is not to be defeated. I do not regard myself as either a victim or a beneficiary of divine dispensations, and I do not believe—much though I should like to do so—that God watches over the welfare of small children in the way that small children need looking after (that is, as fond parents do, and paediatricians and good schoolteachers). I do not believe that God does so because there is no reason to believe it. I suppose that's just my trouble: always wanting reasons.

To abdicate from the rule of reason and substitute for it an authentication of belief by the intentness and degree of conviction with which we hold it can be perilous and destructive. Religious belief gives a spurious spiritual dimension to tribal enmities, as we see them in the Low Countries, Ceylon, Northern Ireland and parts of Africa; nor has any religious belief been held with greater passion or degree of conviction than the metaphysics of blood and soil which did so much to animate Hitler's Germany. Was that not also a consequence of just such a deep, passionate conviction as that which has been thought to authenticate religious belief?

The problem of pain has not been solved, though it has been almost hidden from view by a cloud of theological humbug and the still greater exertions of doublethink that conceal from view or pretend the nonexistence of the most unwelcome truth of all. It goes with the passionate intensity and deep conviction of the truth of a religious belief, and of course of the importance of the superstitious observances that go with it, that we should want others to share it—and the only certain way to cause a religious belief to be held by everyone is to liquidate nonbelievers. The price in blood and tears that mankind generally has had to pay for the comfort and spiritual refreshment that religion has brought to a few has been too great to justify our entrusting moral account-ancy to religious belief. By 'moral accountancy' I mean the judge-ment that such and such an action is right or wrong, or such a man good and such another evil.

I am a rationalist—something of a period piece nowadays, I admit—but I am usually reluctant to declare myself to be so because of the widespread misunderstanding or neglect of the distinction that must always be drawn in philosophic discussion between the *sufficient* and the *necessary*. I do not believe—indeed, I deem it a comic blunder to believe—that the exercise of reason is *sufficient* to explain our condition and where necessary to rem-edy it, but I do believe that the exercise of reason is at all times unconditionally *necessary* and that we disregard it at our peril. I and my kind believe that the world can be made a better place to live in—believe, indeed, that it has already been made so by an endeavour in which, in spite of shortcomings which I do not conceal, natural science has played an important part, of which my fellow scientists and I are immensely proud. I fear that we may never be able to answer those questions about first and last things that have been the subject of this short essay—questions to do with the origin, purpose and destiny of man; we know, however, that whether as individuals or as political people, we do have some say in what comes next, so what could our destiny be except what we make it?

To people of sanguine temperament, the thought that this is so is a source of strength and the energizing force of a just and honourable ambition.

The dismay that may be aroused by our inability to answer questions about first and last things is something for which ordinary people have long since worked out for themselves Voltaire's remedy: 'We must cultivate our garden.'

20 On living a bit longer

I am one of those who believes that a good life—one that is worth living—might well last a bit longer than it normally does. I attach a table compiled mainly from Swedish data showing the mean expectation of life of men and women at various ages over a 200-year period. Some of the variables that affect life expectancy are approaching limiting values—among them mortality at or around birth. Clearly there is room for improvement about expectancy towards life's end. However, most discussion of the matter has been obfuscated by the condemnation of such an ambition as impious and socially destructive, as if that could settle the matter once and for all.

Table: Mean Expectation of Life of Males and Females in Sweden over Two Centuries

Period		Mean Expectation at Age		
		0	60	80
1755–76	Male	33.20	12.24	4.27
	Female	35.70	13.08	4.47
1856–60	Male	40.48	13.12	3.12
	Female	44.15	14.04	4.91
1936–40	Male	64.30	16.35	5.25
	Female	66.90	17.19	5.49
1971–5	Male	72.07	17.65	6.08
	Female	77.65	21.29	7.28

We can witness already the first stirrings of research of which the end result may be the prolongation of active life by the consumption of antoxidant substances. I cannot summarize the matter more briefly than in the essay I wrote for the 'Futures' section of the *Guardian* under the heading 'Four Score Years and Ten—and Still Counting' (13 December 1984). What follows is a slightly edited version of that essay.

Four Score Years and Ten—and Still Counting

It is the great glory and also the great threat of science that anything which is possible in principle—which does not flout a bedrock law of physics—can be done if the intention to do it is sufficiently resolute and long sustained. If therefore a scientific enterprise threatens to endanger or radically to alter our style of life it should be subjected to political scrutiny before being embarked upon—I mean a scrutiny from outside science that gives more weight to moral, social, and prudential considerations than scientists ordinarily give them. Consider in this light the prevalence of the notion that our span of active life might well be extended. Over the past few years the advocacy of Linus Pauling and the experiments of Denham Harman of the University of Nebraska have shown that the lives of laboratory animals can be extended 20–25 per cent by the administration of fairly high doses of substances related to industrial antoxidants. There is no doubt about the authenticity of the experiments: one such was done under my nose by Alex Comfort and his colleagues I. Youhotsky-Gore and K. Pathmanathan in the Zoology Department of University College London. The interpretation of these findings is ambiguous. The mice thus fed will indeed live longer—but is this, as Denham Harman thinks, because antoxidants annul the otherwise destructive action of free radicles on very small biological structures, or could it be because antoxidants are so disagreeable that they seriously put mice off their wittles—thus in effect reproducing the classic experiments by McCay and his colleagues which showed that calorie starvation effectively prolongs life? If these findings in mice and rats could be reproduced in human beings, their effect would be that a man of ninety would have the same energy and address as a seventy-year-old—and so proportionately at other ages; then people would begin to think of our allotted life as *four* score years and ten.

Some professional or amateur gerontologists are keen to try to reproduce their laboratory findings in human beings. Now is this an irreverent and foolhardy escapade which will surely have evil consequences or a bold and exciting scientific adventure of the kind Sir Francis Bacon would have applauded? Consider first the charge of irreverence: it was not God who said that our lifespan was three score years and ten, it was a poet (Psalm 90: 10). Poets are sometimes more influenced by rhyme and metre than by empirical truth or even sense. So it was, for example, with John Dryden in the lines that helped to perpetuate the gothic illusion of a close connection between genius and insanity:

> Great wits are sure to madness near alli'd,
> And thin partitions do their bounds divide.

Even the psalmist did not say he was *sure*.

People temperamentally opposed to attempts to prolong life are fond of quoting Walter Savage Landor's

> Nature I loved, and next to Nature, Art;
> I warmed both hands before the fire of life;
> It sinks, and I am ready to depart.

In Aldous Huxley's *Crome Yellow* a young man taking leave of his hosts taps the barometer in the entrance hall, sees it fall and says—hoping to be overheard and counted as a wit—'It sinks and I am ready to depart.' This is the best comment known to me on Landor's spiritless affirmation.

Healthy and cared-for people have reasons enough to live a few more years—to see how the grandchildren turn out and if the unfolding of history corroborates or confutes their expectations of the way things will go; and a gardener who may have just been replanting a south-facing bed will surely want to gratify the joyous expectation of another spring.

Is the attempt to prolong human life a premeditated insult to nature—an attempt to substitute the inmates of a geriatric ward with the bounding, exuberant, yea-saying folk we were when the world was young? People weren't, of course. Much nearer the truth was Thomas Hobbes's belief that before the coming of Leviathan, that great organism of State, the life of man was solitary, poor, nasty, brutish, and short—and as a rule, ailing too. The 'good old days' argument cuts no ice in medical circles. The time will come when we look back upon the ideology of three score years and ten much as we look back now on the days when a woman would have to bear seven to ten children in the hope of bringing four or five to adult life.

The prolongation of a good life, happy and healthy, is fully in keeping with the spirit of medicine and is in a sense the very consummation of all that medical research has worked towards, *for all advances in medicine increase life-expectancy.* Even a couple of aspirin tablets taken daily might circumvent a platelet crisis and so be seen on an epidemiological scale to increase life-expectancy. The same would go for putting a plaster on a cut finger, something that will infinitesimally reduce the chances of septicaemia, and so on. The prolongation of life will increase population size at a time when there are enough people in the world already, and although the people added will be post-reproductive in age they will still eat and occupy space and consume energy. A graver problem is the burden upon a caring State of pensions and medical care, a burden falling disproportionately on the young. Moreover, working years and provisions for pensions will have to change. These are grave problems but they are not insoluble, for social changes of an essentially similar kind have happened over the past two hundred years, during which the mean expectation of life rose from about 30 to between 60 and 70, and anyhow the changes are not going to take place overnight. Consider the changes since Jane Austen's day. In her first novel *Sense and Sensibility* the elderly Colonel Brandon seeks the hand of a romantic young girl, Marianne Dashwood. In Marianne's view he is an old man who should be thinking not of matrimony but of woolly underwear and how best to avoid draughts. How old then was this amorous old dotard? He was just over thirty-five, we read; and when the question arises of purchasing an annuity for Marianne's mother, a 'healthy woman of forty', it is thought most unlikely that she will live until the age of fifty-five. Suppose now that some prescient man were to have told Jane Austen's characters that over the next century the mean expectation of life at all ages would double—something much more far reaching than the modest 20–25 per cent we now have here in mind—would not Jane Austen's characters have been very mistaken to have been shocked by the riskiness and impiety of such a possibility? The great social adaptation was, however, made and there is no reason to think it cannot be made again. Let us take care that people as far distant from us as we are from the world of Jane Austen do not have reason to pity us for being so faint spirited. But there is an element of risk: we cannot foresee all the distant consequences of increasing life-expectancy, especially in respect of the risk of cancer and perhaps of Alzheimer's disease, senile dementia, and many may think this element of uncertainty should turn us away from our project. What I believe will happen is that some enthusiasts, especially in California, will go ahead with the longevity project to purchase extra years of life at the risk of contracting senile dementia and for its own reward. Francis Bacon, though pious and deeply religious, was the first advocate of just this kind of adventurousness in science, and would have approved: 'The true aim of Science', he wrote in his little-known *Valerius Terminus*, 'is the discovery of all operations and all possibilities of operations from immortality (if it were possible) to the meanest mechanical practice.'

Bacon, then, was on my side. We already have a moral commitment to biomedical research which increases life-expectancy and I see no reason to think that the highway of medical melioration that has brought us so far already will now lead us into evil. We have long since been travelling that road and it is too late now to cease to be ambitious.

This is all very fine, but we have no scientific authority either to believe or to doubt that the results of these laboratory experiments on the longevity of animals are applicable to man. Inspired by these findings our own practice is to take 5 g of ascorbic acid (vitamin C) every day, and 2,000 IU of vitamin E—an antioxidant widely used in manufactures to prevent the rancidification of fats, especially vegetable oils. Gestures such as these will be thought timid by American colleagues who take 10 g a day and excessive by people naïve enough to suppose that the metabolic function of vitamin C in nature is to prevent scurvy. It may be that we have by luck hit upon a good compromise, but whether we have or not I have no intention of preparing any last words, partly because I do not expect to be in a garrulous mood and partly because the utterance of last words is a deeply unsatisfactory art form. I am certain though that my last *thoughts* will be of Jean and that, so far as they concern my life generally, I shall be thinking that in spite of its vicissitudes my life has by no means been without its risible aspects.

Notes

Chapter 1 *The Phenomenon of Man*

1. (London, 1959).
2. p. 153.
3. p. 16 and again p. 18; p. 19.
4. pp. 12, 13.

Chapter 2 *Hypothesis and imagination*

1. *Hypotheses non sequor,* runs an early draft of Newton's famous disclaimer, which we are to translate, as Whewell did, 'I feign no
   hypotheses': see I. Bernard Cohen, *Isis,* **51,** 589, 1960. Newton did, of
   course, use and propound hypotheses in the modern sense of that
   word; the unwholesome flavour which Newton found in the word is
   discussed below.
2. The fundamental axiom of empiricism—*nihil in intellectu quod non
   prius in sensu*—is of course mistaken. Animals *inherit* information (for
   example, on how to build nests, or what to sing) in the form of a sort
   of chromosomal tape-recording. This instinctual knowledge is not
   arrived at by association of ideas, anyhow of sensory ideas received by
   the animal in its own lifetime.
3. *The Organization of Behavior* by D. O. Hebb (New York, 1949), especially p. 31.
4. Here and hereafter I quote from the following works of the authors
   cited under heading 4 in the text: *The Philosophy of the Inductive
   Sciences,* by William Whewell, in 2 vols, 2nd ed. (London, 1847; 1st
   ed., 1840); *The Principles of Science,* by W. Stanley Jevons, 2nd ed.,
   revised (London, 1877; 1st ed., 1873); *The Art of Reasoning,* by Samuel
   Neil: twenty articles in successive issues of the first two vols of the
   British Controversialist (1850, 1851), of which Neil was editor; particularly vol. 2, no. 11, *Collected Papers* of C. S. Peirce, ed. C. Hartshorne
   and P. Weiss, vol. 2, *Elements of Logic* (Cambridge, Mass., 1932), vol.
   6, *Scientific Metaphysics* (1935).

5. *Elements of the Philosophy of the Human Mind*, by Dugald Stewart, 2nd ed., 2 vols (London, 1802, 1816; 1st ed., 1792, 1814).

6. *The Principles of Empirical or Inductive Logic*, 2nd ed. (London, 1907; 1st ed., 1889).

7. *A System of Logic*, by John Stuart Mill, 8th ed. (London, 1872; 1st ed., 1843).

8. *Elements of Logic*, by Richard Whately, 9th ed. (London, 1848; 1st ed., 1826).

9. *Charles Darwin*, by Gavin de Beer (London, 1963), p. 98. In his autobiography Darwin once declared that he could not resist forming a hypothesis on every subject, and his letters to Henry Fawcett and to H. W. Bates are very revealing: *More Letters of Charles Darwin*, ed. F. Darwin and A. C. Seward (London, 1903), pp. 176, 195.

For the origin of the idea of Natural Selection, see also L. Eisely in *Daedalus*, Summer 1965, pp. 588–602.

10. *The Grammar of Science*, by Karl Pearson, 3rd ed. (London, 1911; 1st ed., 1892).

11. e.g. 'science', 'art', 'pure science', 'applied science', 'analysis', 'synthesis', 'experiment'; and, of course, notoriously, words like 'genius', 'creation', 'enthusiasm'.

12. *On the Duties and Qualifications of a Physician*, new ed. (London, 1820, 1st ed., 1772).

13. *Essays on the Intellectual Powers of Man*, 1st ed., 1785. In *The Works of Thomas Reid, D.D.*, ed. W. Hamilton, 4th ed. (London, 1854).

14. *Discourse on the Study of Natural Philosophy* (London, 1831).

15. *Problems of Life and Mind*, 4th ed. (London, 1883; 1st ed. 1873), esp. pp. 296, 316–17.

16. *A System of Logic*, book III, chapter 14, § 6.

17. See Stephen Hale's Preface to his *Statical Essays*, 4th ed. (London, 1769; 1st ed., 1727) and a number of passages in Robert Hooke's *Posthumous Works* (London, 1705). For Boscovich, see note 20.

18. *On Method*, by Samuel Taylor Coleridge, 3rd ed. (London, 1849; 1st ed., 1818).

19. *De augmentis scientiarum*, trans. Gilbert Wats (London, 1674), book 5, chapter 3, II.

20. Dugald Stewart's translation of the footnotes on pp. 211–12 of Boscovich's *De solis ac lunae defectibus* (London, 1760).

21. *The Design of Experiments*, by R. A. Fisher (Edinburgh, 1935).

22. *Observations on Man*, vol. 1 (London, 1749), pp. 15–16.

23. In particular see book 1, chapter 2, §§ 9, 10 (2nd ed.).

24. Robert Adamson in his article 'Bacon, Francis' in the 9th ed. of the *Encyclopaedia Britannica* (Edinburgh, 1875). See also Augustus de Morgan's *A Budget of Paradoxes* (London, 1872): 'Modern discoveries have not been made by large collections of facts . . . A few facts have

suggested an *hypothesis*, which means a *supposition*, proper to explain them. The necessary results of this supposition are worked out, and then, and not till then, other facts are examined to see if these ulterior results are found in nature. The trial of the hypothesis is the *special object* . . . Wrong hypotheses, rightly worked from, have produced more useful results than unguided observation.'

25. See Popper's 'Science: Problems, Aims, Responsibilities', in *Federation Proceedings* (Federation of American Societies for Experimental Biology), **22**, 961–72, 1963, and my own broadcast 'Is the Scientific Paper a Fraud?' (Chapter 3 of this volume). This broadcast was followed by a correspondence (*Listener*, 26 September and 10 October, 1963) illustrating the style of thought that makes scientists treat the 'philosophy of science' with exasperated contempt.

26. *Introduction à l'étude de la médicine expérimentale* (Paris, 1865).

27. See for example J. Bronowski, *Science and Human Values*, revised ed. (New York, 1965).

Chapter 4 *The Act of Creation*

1. (London, 1964).
2. For example, his account (p. 452) of C. M. Child's explanations of 'physiological isolation'.
3. 'Noise' as the word is used by communication theorists: I borrowed this thought from an impromptu of Professor A. L. Hodgkin's.
4. 'Hypothesis and Imagination', Chapter 2 of this volume.
5. pp. 94–5.
6. (London, 1982).

Chapter 5 *Darwin's illness*

1. *New England Medical Journal*, **261**, 1109, 1959.
2. *Lancet*, **244**, 129, 1943; **265**, 1351, 1953.
3. C. D. Darlington, *Darwin's Place in History* (Oxford, 1959).
4. *The Times*, 31 December 1963. For Kempf, see *Psychoanalytic Review*, **5**, 151, 1918.
5. *Lancet*, **1**, 106, 1954.
6. *The Quest for the Father* (New York, 1963).
7. *Nature*, **184**, 1102, 1959.
8. (London, 1963).
9. In this connection, however, my friend Professor P. C. C. Garnham FRS writes: 'The initial infection in many cases is probably unaccompanied by acute symptoms: there is often no more than a small "insect bite" which goes unnoticed amongst all the others, while this inconspicuous lesion is followed by a long period of latency, with no symptoms, until the person falls down dead at the age of 40 or 50

with a ruptured aneurysm; in fact one endemic focus in Brazil is known as the "Land of Sudden Death". Sometimes the course is less dramatic, and, as with Darwin, a chronic illness arises, with signs and symptoms so insidious that the correct diagnosis is often missed.'

Chapter 6   *Two conceptions of science*

1. This is one (by no means the longest) of eight questions set by Professor Grant in Comparative Anatomy in February 1860:

'By what special structures are bats enabled to fly through the air? and how do the galeopitheci, the pteromys, the petaurus, and petauristae support themselves in that light element? Compare the structure of the wing of the bat with that of the bird, and with that of the extinct pterodactyl: and explain the structures by which the cobra expands its neck, and the saurian dragon flies through the atmosphere. By what structures do serpents spring from the ground, and fishes and cephalopods leap on deck from the waters? and how do flying-fishes support themselves in the air? Explain the origin, the nature, the mode of construction, and the uses of the fibrous parachutes of arachnidans and larvae, and the cocoons which envelope the young; and describe the skeletal elements which support, and the muscles which move the mesoptera and the metaptera of insects. Describe the structure, the attachments, and the principal varieties of form of the legs of insects; and compare them with the hollow articulated limbs of nereides, and the tubular feet of lumbrici. How are the muscles disposed which move the solid setae of stylaria, the cutaneous investment of ascaris, the tubular peducle of pentalasmis, the wheels of rotifera, the feet of asterias, the mantle of medusae, and the tubular tentacles of actinae? How do entozoa effect the migrations necessary to their development and metamorphoses? how do the fixed polypifera and porifera distribute their progeny over the ocean? and lastly, how do the microscopic indestructible protozoa spread from lake to lake over the globe?'

2. This not a debating point, for Shelley's writing can sustain both views. In his *Defence of Poetry* Shelley defines poetry in a 'universal' sense that comprehends all forms of order and beauty, and includes, therfore, not merely poetry in the narrower sense, but science as well (poetry 'comprehends all science'). Earlier, however, Shelley put Reason and Imagination at opposite poles; if then, as in the second conception I outline, science is regarded as an essentially rational activity, Shelley may quite rightly be allowed to speak for the view that science and poetry are antithetical. See D. G. King-Hele in the *New Scientist* **14**, 352–4, 1962; and Graham Wallas, *The Art of Thought* (London, 1926).

3. For the conception that *truth is manifest*, see the critical analysis by Karl Popper, 'On the Sources of Knowledge and of Ignorance', in *Conjectures and Refutations* (London, 1963). The question where Truth resides can also be put of Beauty, and answered in the same two ways, for the romantic view does not distinguish them. For the history of the idea that *beauty is manifest* (as opposed to being in the eye of the observer), see Logan Pearsall Smith, *The Romantic History of Four Words: romantic, originality, creative, genius*, S.P.E. Tract 17 (Oxford, 1924).

4. For earlier accounts of the hypothetico-deductive scheme, see 'Hypothesis and Imagination', Chapter 2 of this volume.

5. *The Great Instauration*, Preface.

6. V. B. Wigglesworth has pointed out that in the first edition of his *Grammar of Science* (1892) Karl Pearson chose Hertzian waves, essentially radio waves, as an example of a discovery of no apparent usefulness. This is the best example I know of the apparently useless bringing in the goods (as Pearson thought it probably would).

7. Thomas Sprat, *History of the Royal Society* (1667), p. 245.

8. Benjamin Britten always spoke in favour of occasional music: see his speech *On Receiving the first Aspen Award* (London, 1964).

9. See M. Goldsmith and A. L. Mackay (eds.), *The Science of Science* (London, 1964).

10. A 'story' is more than a hypothesis: it is a theory, a hypothesis together with what follows from it and goes with it, and it has the clear connotation of completeness within its own limits. I notice that laboratory jargon follows this usage, e.g. 'Let's get So-and-so to tell his story about' something or other, an invitation which So-and-so may decline on the grounds that his work 'doesn't make a story yet' or accept because he 'thinks he's got a story'. There is a slightly depreciatory flavour about this use of 'story' because fancy has to be used to fill in the gaps and some people tend to overdo it.

Chapter 7   *Science and the sanctity of life*

1. An eminent theologian once said to me that I was making altogether too much fuss about this kind of mischance. It was, he said, all in the nature of things and already comprehended within our way of thinking; it was not different in principle from being accidentally struck on the head by a falling roof tile. But I think there *is* an important difference of principle. In the process by which a chromosome is allotted to one germ cell rather than another, and in the union of germ cells, luck is of the very essence. The random element is an integral, indeed a defining, characteristic of Mendelian inheritance. All I am saying is that it is difficult to wear a pious expression when

the fall of the dice produces a child that is structurally or biochemically crippled from birth or conception.
2. There are, perhaps, weighty legal and social reasons why even tragically deformed children should be kept alive (for who is to decide? and where do we draw the line?), but these are outside my terms of reference.
3. See my book *The Future of Man* (New York, 1959; London, 1960).
4. These arguments are set out more fully in my 'The Genetic Improvement of Man' in *The Hope of Progress* (London, 1972), pp. 69–76.
5. See Stephen Toulmin and June Goodfield, *The Discovery of Time* (London and New York, 1965).
6. 'We are almost the last progeny of the First Men,' said Thomas Burnet.

Chapter 8   *J.B.S.*

1. *J.B.S.: The Life and Work of J. B. S. Haldane* (London, 1968).

Chapter 9   *Lucky Jim*

1. W. H. Auden, *A Certain World* (London, 1971).

Chapter 10   *On 'the effecting of all things possible'*

1. See, for example, Herbert Grierson, *Cross Currents in English Literature of the Seventeenth Century* (London, 1929); Basil Willey, *The Seventeenth-Century Background* (London, 1934); G. N. Clark, *The Seventeenth Century*, 2nd ed. (Oxford, 1947); W. Notestein, *The English People on the Eve of Colonization* (New York, 1954); Marjorie Hope Nicolson, *Mountain Gloom and Mountain Glory* (Ithaca, 1959); Maurice Ashley, *England in the Seventeenth Century*, 3rd ed. (London, 1961); H. R. Trevor-Roper, *Religion, the Reformation and Social Change* (London, 1967).
2. George Williamson, 'Mutability, Decay and Seventeenth Century Melancholy', *Journal of English Literature and History*, **2**, 121–50, 1935.
3. Christopher Hill, *Intellectual Origins of the English Revolution* (Oxford, 1965).
4. Marjorie Hope Nicolson, 'Two Voices: Science and Literature', *Rockefeller Review*, **3**, 1–11, 1963.
5. See, for example, Margery Purver, *The Royal Society: Concept and Creation* (London, 1967) and a number of papers in *Notes and Records of the Royal Society*, **23**, no. 2, December 1968.
6. For England in particular, see Christopher Hill, op. cit. (note 3); F. R. Johnson, *Astronomical Thought in Renaissance Britain* (Baltimore, 1937).

7. England at the time of the Armada was a prosperous country, and it became so again in the reign of Queen Anne; the period I am discussing, however, was marked by a high level of unemployment and a number of major economic slumps, not to mention the English Civil War; moreover the reputation of England abroad sank to a specially low level in the latter part of James I's reign and during the reign of Charles I. This was also the period of the great emigrations to Massachusetts.

8. William Lecky, *The Rise and Influence of Rationalism in Europe* (London, 1865); see especially H. R. Trevor-Roper, op. cit. (note 1).

9. J. V. Andreae's, *Description of the Republic of Christianopolis* was first published in 1619—see F. Held, *Christianopolis, an Ideal State of the Seventeenth Century* (Urbana, 1914); Tommaso Campanella published *The City of the Sun* in 1623—English translation by T. W. Halliday in *Ideal Commonwealths* (London, 1885). There is an extensive literature on Utopian and chiliastic speculation, some of it rather feeble. The following are specially relevant to the idea of progress and of human improvement: J. B. Bury, *The Idea of Progress* (London, 1932); E. L. Tuveson, *Millennium and Utopia* (Berkeley, 1949); Norman Cohn, *The Pursuit of the Millennium* (New York, 1957).

10. In a sermon delivered in Whitehall, 24 February 1625.

11. In *Hydriotaphia*, his discourse on urn-burial. For a history of the idea of time, see Toulmin and Goodfield, op. cit. (Chapter 7, note 5).

12. *An Apologie of the Power and Providence of God* (Oxford, 1627), an answer to Godfrey Goodman's *The Fall of Man* (London, 1616).

13. See D. C. Allen, 'The Degeneration of Man and Renaissance Pessimism', *Studies in Philology*, **35**, 202–27, 1938.

14. Louis le Roy's remarkable work, addressed to 'all men who thinke that the future belongeth unto them', became known in England through Robert Ashley's translation of 1594 (*Of the Interchangeable Course or Variety of Things*).

15. Cited by Hakewill, op. cit. (note 12).

16. *Experimental Philosophy* (London, 1664), Preface; cf. John Ray, *The Wisdom of God Manifested in the Works of the Creation* (London, 1691), pp. 124–5.

17. William Godwin, *An Enquiry Concerning Political Justice*, 3rd ed. (London, 1797; 1st ed., 1793), vol. I, p. xxvi; vol. II, p. 535. Cf. Adam Ferguson, *An Essay on the History of Civil Society* (Edinburgh, 1767).

18. The idea of 'stretched experience', and of the experimenter as the 'archmaster' who 'completes experience', comes from John Dee's 'Mathematical Preface' to Henry Billingsley's English translation of Euclid (London, 1570).

19. C. Day Lewis, *A Hope for Poetry* (London, 1934), p. 107.

20. This simile occurs more than once in Hobbes; the passage I have in mind is from is *Human Nature* (London, 1650), chapter 9.

Chapter 11   *Further comments on psychoanalysis*

1. In mathematics, x is a function of y when the value of x varies in dependence on the value of y.
2. To speak (as I do here and below) of the causes of *differences* between human beings sounds clumsy and takes some getting used to; but there seems to be no avoiding it if one is to be precise and at the same time avoid a formal symbolic treatment.
3. *Conjectures and Refutations* (London, 1963), pp. 34–9.
4. A disease of the kind psychoanalysts would be well advised not to meddle with.
5. 'Curing is so ambiguous a term', says Dr David Cooper in *Psychiatry and Anti-Psychiatry*; 'one may cure bacon, hides, rubber, or patients. Curing usually implies the chemical treatment of raw materials so that they may taste better, be more useful, or last longer. Curing is essentially a mechanistic perversion of medical ideals that is quite opposite in many ways to the authentic tradition of healing.' Somewhat similar views are to be found in the writings of R. D. Laing, Michel Foucault and J. Lacan.
6. See B. A. Farrell, introduction to *Leonardo da Vinci* (Harmondsworth, 1963).
7. 'Darwin's illness', Chapter 5 of this volume.

Chapter 12   *The strange case of the spotted mice*

1. Joseph Hixson, *The Patchwork Mouse* (Anchor Press, 1976).
2. Dr Summerlin was not in fact funded by the National Institutes of Health nor other benefactors, but from Dr Good's discretionary funds—a fact pointed out by Mr Hixson and Dr Robert S. Schwartz in the *New York Review*, 10 June 1976, and accepted by Sir Peter.
3. *The Art of the Soluble* (London, 1967).

Chapter 13   *Unnatural science*

1. 'Fifty Years' Progression in Soil Physics', *Geoderma*, 12, 265–80, 1974.
2. *American Journal of Human Genetics*, 28, 107–22, 1976.
3. (Potomac, Md, 1974).
4. Eysenck developed his interpretation in *Encounter* for January 1977 and says that the most Gillie was entitled to say was that 'there were certain inconsistencies in Burt's data which called in question the interpretation which might be put upon them'.

5. Ned Block and Gerald Dworkin (eds.), *The IQ Controversy: Critical Readings* (New York, 1976; London, 1977).

Chapter 14 *Florey story*

1. Gwyn Macfarlane, *Howard Florey: The Making of a Scientist* (Oxford, 1979).

Chapter 15 *In defence of doctors*

1. In *The Role of Medicine* (Princeton, 1980).
2. Ivan Illich.

Chapter 16 *Expectation and prediction*

1. For example, it is generally admitted to be true that Germany would not have been able to embark upon the First World War unless Fritz Haber had worked out how to 'fix' atmospheric nitrogen and turn it into nitrates or compounds of ammonia, so making Germany independent of imported fertilizers. This is the best example I know of how advances in technology can influence the course of history.
2. See 'Unnatural science', Chapter 13 of this volume.
3. See 'Lucky Jim', Chapter 9 of this volume.

Chapter 17 *Scientific fraud*

1. W. Broad and N. Wade, *Betrayers of Truth: Fraud and Deceit in the Halls of Science* (London, 1983).
2. The Summerlin case is omitted here because there is a much fuller account in Chapter 12.

Chapter 18 *Son of stroke*

1. From Sir Peter's introductory note to this light-hearted article (published in *World Medicine*, 18 Oct. 1972):

   he [the author, i.e. himself] threatened to write a book of memoirs to be called 'Stroke', hoping by this means to secure specially indulgent attention from any members of the nursing staff who wished to cut a good figure in his pages. When his attention was called to the fact that somebody has already written a book called 'Stroke', he altered his title to 'Son of Stroke'.

   The paragraphs that follow represent the precious distillate of this major literary work.

Chapter 19   *The question of the existence of God*

1. *The Limits of Science* (New York, 1984; Oxford, 1985).
2. *Francis Bacon: From Magic to Science*, trans. S. Rabinovitch (London, 1968).

Sources

This selection is taken from the published writings of Peter Medawar. The following list gives details of the essays' first publication. Most were later published in the two collections, *Pluto's Republic* and *The Threat and the Glory*.

1. *The Phenomenon of Man*

Review of Pierre Teilhard de Chardin, *The Phenomenon of Man*, *Mind* **70** (1961).

2. *Hypothesis and imagination*

Times Literary Supplement, 25 October 1963; reprinted in expanded form in *The Art of the Soluble*.

3. *Is the scientific paper a fraud?*

Unscripted broadcast on BBC Third Programme, *Listener* **70**, 12 September 1963. Reprinted in *The Threat and the Glory*; by permission of HarperCollins, New York.

4. *The Act of Creation*

Review of Arthur Koestler, *The Act of Creation*, *New Statesman*, 19 June 1964.

5. *Darwin's illness*

Review of Phyllis Greenacre, *The Quest for the Father*, and Gavin de Beer, *Charles Darwin*, *New Statesman*, 3 April 1964.

6. *Two conceptions of science*

Henry Tizard Memorial Lecture, *Encounter* **143** (August 1965).

7. *Science and the sanctity of life*

Encounter **27**, no. 6 (December 1966).

8. *J.B.S.*

Review of Ronald Clark, *J.B.S.: Life and Work of J. B. S. Haldane*, *New York Review of Books*, 10 October 1968.

9. Lucky Jim

Review of J. D. Watson, *The Double Helix, New York Review of Books*, 28 March 1968.

10. On 'the effecting of all things possible'

Presidential address to the British Association, 1969, in *The Hope of Progress.*

11. Further comments on psychoanalysis

From *The Hope of Progress.*

12. The strange case of the spotted mice

Review of Joseph Hixson, *The Patchwork Mouse* (Anchor Press, 1976), *New York Review of Books*, 15 April 1976. Reprinted in *The Threat and the Glory*; by permission of HarperCollins, New York.

13. Unnatural science

Review of Leon J. Kamin, *The Science and Politics of IQ*, and N. J. Block and Gerald Dworkin (eds.), *The IQ Controversy, New York Review of Books*, 3 February 1977.

14. Florey story

Review of Gwyn Macfarlane, *Howard Florey: The Making of a Scientist* (Oxford University Press, 1979), *London Review of Books*, 20 December 1979. Reprinted in *The Threat and the Glory*; by permission of HarperCollins, New York.

15. In defence of doctors

Review of Thomas McKeown, *The Role of Medicine* (Princeton, 1980), *New York Review of Books*, 15 May 1980. Reprinted in *The Threat and the Glory*; by permission of HarperCollins, New York.

16. Expectation and prediction

From *Pluto's Republic.*

17. Scientific fraud

Review of W. Broad and N. Wade, *Betrayers of Truth: Fraud and Deceit in the Halls of Science* (Century, 1983), *London Review of Books*, 17–30 November 1983. Reprinted in *The Threat and the Glory*; by permission of HarperCollins, New York.

18. Son of stroke

World Medicine, 18 October 1972. Reprinted in *The Threat and the Glory*; by permission of HarperCollins, New York.

19. *The question of the existence of God*
From *The Limits of Science*. By permission of HarperCollins, New York.

20. *On living a bit longer*
From *Memoir of a Thinking Radish*.

Books by Peter Medawar

The Uniqueness of the Individual (London, 1957)
The Future of Man: BBC Reith Lectures 1959 (London, 1960)
The Art of the Soluble (London, 1967)
Induction and Intuition in Scientific Thought (London, 1969)
The Hope of Progress (London, 1972)
Advice to a Young Scientist (New York, 1979)
Pluto's Republic (Oxford, 1982)
The Limits of Science (New York, 1984)
Memoir of a Thinking Radish (Oxford, 1986)
The Threat and the Glory (Oxford and New York, 1990)

with Jean Medawar

The Life Science (London, 1977)
Aristotle to Zoos (Cambridge, Mass., 1983)

Index